Faith
Enterprise

Faith Enterprise

Making Mysteries and Miracles Vanish

Richard Maus

ANTERIOR PUBLISHING
Northfield, Minnesota

Published by:
ANTERIOR PUBLISHING
204 West 7th Street, Box 8
Northfield, MN 55057

For information on bulk sales or academic textbook sales
contact the publisher.
faithenterprise.info

Cover design by: Amy Kirkpatrick, kirkpatrickdesign.com
Edited by: Nancy Ashmore, njashmore@gmail.com
Interior design by: Dorie McClelland, springbookdesign.com

Pre-Release Edition, July 2016
Printed in the United States of America
21 20 19 18 17 16 2 3 4 5 6 7 8

ISBN: 978-0-9776205-2-4 (hardcover)
ISBN: 978-0-9776205-3-1 (softcover) Release date – November 22, 2016

Contents

Jews do not recognize Christ.

Protestants do not recognize the Pope.

And Baptist friends don't
recognize each other . . .
when they meet in a liquor store.

(Source unknown)

Prologue

Throughout this book we will assume the existence of a Supreme Being, a God who has the following four characteristics. We will refer to these as the *God Premises.*

THE GOD PREMISES

God is omnipotent—has unlimited power and authority

God is omniscient—has complete knowledge of everything

God is omnipresent—is present everywhere at the same time

God is omnibenevolent—is perfectly good, just, and merciful

These premises are regularly associated with *mysteries*, which we will also explore.

Introduction

I'm far from the only person to have wrestled with questions about religion. People have been doing it as long as religion has existed.

Two hundred years ago, Thomas Paine wrote in *The Age of Reason* that the Benedictine understanding of angels was obsolete—and had been for four hundred years before that. Some one hundred years ago, Mark Twain wrote in *Letters from the Earth* that *No prayer of [humankind's] has ever been answered.* Mother Teresa spoke of an agony and angst about faith and not seeing evidence of Jesus that was so profound that it caused her to stop praying. George W. Bush, on the other hand, declared, *I believe that God wants me to be President.*

If such people can have convictions or questions about faith and Jesus, anyone can, including a retired teacher looking for answers to questions he first encountered in grade school.

My first fourteen years of formal education were guided by Benedictine priests, monks, and nuns. After completing my study of mathematics and physics at public universities, I taught for three decades in Minnesota public schools. This early parochial, religious education followed by a secular education in science has given me a unique vantage point—and generated many questions, questions that I have researched at length and attempt to expand on in this book.

Discourse relating to faith, science, and religion can be esoteric, convoluted, fuzzy, emotional, subjective, or opinionated. I have tried to make this book simple, factual, understandable, and accessible to all. I want readers to take their time reading it and to compare what it says with the precepts and scriptures of their faiths or creeds. I want them to ponder what they discover in the process.

As I explore this subject with readers, I am relying on two things: 1) A reader with a mind open to seeing and exploring common-sense observations based on reason, as well as on the revealed word of God, and 2) information that we can examine and verify today. I have documented many of my sources, all but those for which information is readily available in public libraries or via online search engines such as Bing or Google. (They should be rare, but if you find errors, please contact me so that I can correct them in future editions of the book.)

A few more words about sources: When I quote the Bible, it will generally be the King James Version, which is the most widely read version and which, since it dates back to 1612 CE, is without copyright restrictions. (You are free to check what is printed in here with the passages in the Bible translation you prefer. If you find that they are significantly different there, make note of those differences—and think about the fact that the word of God can vary from Bible to Bible.)

I allow the Catholic Church to speak for itself via its popes and the publications, proclamations, and organizations officially charged with communicating its philosophies, principles, and policies. As I do so, I use the following abbreviations:

(CCC) Catechism of the Catholic Church
(YOUCAT) Youth Catechism of the Catholic Church
(CCL) Catholic Canon Law—1983
(CCT) Catechism of the Council of Trent
(USCCB) US Conference of Catholic Bishops
(VIS) Vatican Information Service

These quotations are used as is, without alteration from the original, except where ellipses have been inserted to indicate material has been omitted or where I have added emphasis—and made note of that modification. Spelling, capitalization, and formatting are otherwise unchanged.

At times I make claims the validity of which I assume should be obvious to everyone, like noting, for instance, that it is not reasonable, fair, or good that all humanity, throughout all time, was condemned to suffer for the actions of Adam and Eve.

At other times, I quote notable people of various kinds. Some of them said what I have wanted to say first and better. Some of them support what I claim critically. Some are offered as clarifying examples. Their words also remind us that the questions we are exploring are not all new ones.

Much of what I have written about pertains to all religions. When I get specific, however, I draw my examples from Christianity and, more particularly, the Roman Catholic religion, because of my study, observations, and experience with it. When I offer sports analogies, I use examples from baseball, again from familiarity.

You are wondering, probably, why I have written this book. Let me reassure you: I have not written it because I am angry at a god or as a lark. I'm not trying to torment or to tempt, either. My motivations reduce to these four points:

This is, in part, my answer to the Benedictines who oversaw my early schooling. I wasn't able to respond adequately to the indoctrination I was receiving then, though I tried. (Under my photo in the 1957 edition of the yearbook of St. Boniface High School, it says, *There are two answers to every question: mine and the teacher's.*)

Second, most religious training takes place in childhood, and in later years many adults find themselves confused or curious about faith. I was, so I spent six years researching its basic tenets. Writing about what I learned in the process has helped me organize my thoughts and reexamine my core beliefs, something I want to share with others who are seeking answers to old questions.

Third, I was intrigued by the challenge that Sam Sawyer of The Jesuit Post blog issued recently when he said, *Today the task is not to answer the arguments of the New Atheism on their own grounds. Rather we should all ask what it is in Christianity that does such an effective job of making New Atheism attractive.*

The tug of war between religious liberty and laws prohibiting discrimination, a long-time challenge for all levels of government in the United States, is only growing more heated. I am particularly concerned about the struggle between faith and science that is occurring worldwide and the dire consequences that could result from extremist views that deny everything from the humanity and intrinsic worth of those of other religions, races, and creeds to the safety and importance of vaccines and the causes and effects of climate change. As a bonus, I have found it refreshing and encouraging to examine the contributions science has made and continues to make to human progress.

I've organized this book in six sections.

Part I clarifies the differences between faith and science. This will provide a base for understanding the parts which follow. It will also help you understand topics such as *teach the controversy,* church-state separation, prayer in public school, and the history of science.

Part II not only explores religion as a church activity, but also examines its recent history, the origins of morality, and current trends.

Part III deals specifically with a popular world religion: Christianity. In it we will examine the characteristics of its God and what its God wants of us.

Part IV gets into the Bible, the foundation of Christianity. Is it believable? Does it make sense? Can it provide what it claims?

Part V examines a particular Christian religion, Roman Catholicism. We examine how it has dealt with the problems in the Bible. How is it supported today?

In **Part VI** we look at an analogous socio-economic phenomenon that doesn't involve a deity. Why? Because a common response to someone questioning anything at the core of Catholicism/Christianity is to ask how anything could become so large and influential if it wasn't supported by God and His word in the Bible.

Steve Green, president of Hobby Lobby, has said, *This nation is in danger because of its ignorance of what God has taught.* (*New York Times*, July 16, 2014)

I agree with him, though not for the same reasons. I believe our country and our world will be better, safer places, if we not only revisit God's teachings and how accounts of them came to be, but also explore the mysteries that exist within and between versions of the word of God as presented in various Bibles.

In the process of this journey of exploration, I intend to give you the opportunity to learn new facts and to draw some conclusions different from the ones you may hold now. Your only prerequisite is an interest in faith, science, reason, religion, baseball, and the relationships between them—and the belief that as a human being you have a vested interest in a world in which all decisions are rational ones based on supporting evidence.

Part I

Faith and Science

It is impossible to have faith and scientific knowledge about the same thing. (St. Thomas Aquinas)

What is reality? Nearly everybody, young and old, spends time each day trying to answer some form of this question. It might be as simple as *Has the mail been delivered?* or *How long will it take to get to Grandma's house?* or as crucial as *Can I run fast enough to outrun the tiger that's chasing me? How much money will we need for the next payroll?* or *Will we have enough to eat tomorrow?*

Reality is *the totality of all things possessing actuality, existence, or essence* (The American Heritage Dictionary) or that which exists objectively and in fact *within* the everyday natural, physical world we live in. As opposed to that which is called the *supernatural*, things *beyond* the natural world.

Reality may look different to you and me. But it exists. It is something we each experience—from tornadoes, rainbows, and germinating seeds to human-made forms of reality, like IRS regulations and speed limits. Like the capacity of your car's fuel tank.

When you and your friends go for a spin and run out of gas, you are presented with something that you can all observe, verify, and agree upon. The car has stopped, the needle on the gas gauge is pointing to Empty, and you remember you meant to fill the tank yesterday. Endless philosophical debate and denial won't change the situation. This is reality. You ignore it at your own expense. If you want to continue on your way, you need to admit you ran out of gas and begin planning how to get beyond this problem. Handel's *Messiah*, the *Mona Lisa*, novels, and banana splits are real, too. (You ignore these to your own detriment as well.)

Things of this sort are subject to change. Other forms of reality, like the laws of nature, are unchanging, although our understanding of them is not.

Take the night sky, for instance. Deciphering its mysteries has not only taken all of human history to date but will also likely continue as long as the human race survives. In the process, we have learned that the stars never were, are not now, and never will be as we see them today. When we gaze at them, we are looking back in time—a different time for each star. When we look at our nearest star, the sun, we are seeing how it was eight minutes earlier. That bright nighttime star we're admiring might well have disappeared into a black hole or exploded into a wisp of invisible gas millions of years ago. The star that appears to be right next to it might not have come into existence until long afterward.

It's not always easy to recognize or accept this reality. In fact, as we explore the natural world and build our understanding of its realities, it takes a lot of work to avoid delusions and misconceptions.

> *Humankind cannot stand very much reality.* (T. S. Eliot)

In the pages that follow we're going to examine ***faith*** and ***science***, the only two methods available to us in our attempt to acquire the knowledge needed to determine reality. I purposefully want to be very precise in the meaning of these words, so let's look at how they are defined.

Faith (noun): *1. knowing something with no evidence;*
2. a set of religious beliefs, usually relating to the supernatural

We'll examine the second definition at length in Part II. Let's begin with the first definition that is found in many dictionaries and which is used in the Bible: knowing something with no evidence.

The Bible says that *faith is the substance of things hoped for, the evidence of things not seen.* (Hebrews 11:1) St. Augustine, Bishop of Hippo, concurred, saying: *Faith is to believe what you do not see; the reward of this faith is to see what you believe.* (*Sermones* 4.1.1)

The Bible also suggests that God blesses those who believe on faith: *Jesus saith unto him, Thomas, because thou hast seen me, thou hast believed: blessed are they that have not seen, and yet have believed.* (John 20:29)

Mark Twain, on the other hand, wrote that *Faith is believing what you know ain't so.* Contemporary philosopher Peter Boghossian defines it as *pretending to know things you don't know*. Seat of the pants, wishful thinking, gut feeling, and lucky guess—these phrases are all hints you're dealing with faith-based actions.

We often use the word *faith* where there might be evidence encouraging some confidence. When I express my faith that the hummingbirds will return in spring and my lawn will become green again, I'm expressing my confidence in the regularity of nature.

Such confidence is not always well-founded or wise. Consider faith-based driving.

- I back the car out of the garage without looking. I've never had problems before and I have faith that this pattern will continue. This time, however, I run over a bicycle someone left in the driveway.
- While driving near the speed limit in heavy traffic, I encounter a thick fog. I continue driving a car-length or so behind the car in front of me so I can still see its tail lights. If it slows, I will slow. If it stops quickly, I will too. I have no knowledge of what lies ahead, but I have faith that the drivers ahead of and behind me are being careful and conscientious. All is well until reality intrudes upon us with a spectacular 53-car pile-up.

There have been instances of faith-based flying as well. One of the most dramatic was the case of an Air Canada Boeing 767 later called the *Gimli Glider*. No competent pilot—not one flying alone, much less one carrying 61 passengers—would take off without enough fuel to complete the next leg of a trip. The pilots of this July 1983 flight did, however, convinced by printed evidence that sufficient fuel had been loaded into the airplane. The reality was that due to a mistake in metric conversion they had only half the fuel needed to reach their destination. Fortunately they were able to glide to a successful emergency, engines-out landing at Gimli, an old abandoned military airstrip.

> *He who knows nothing is closer to the truth than he whose mind is filled with falsehoods and errors.* (Thomas Jefferson)

We like to think we believe things because of evidence, but our evidence sometimes turns out to be erroneous. This leaves us depending solely on faith. False evidence can be more harmful than no evidence, leading to dangerous delusions:

I'm young, so I can delay saving for retirement; things will work out when the time comes.

It's my *first time,* so we don't need to take precautions; I can't possibly get pregnant.

We don't have to worry about controlling the size of our family; we'll somehow be able to care for, raise, and educate all the children we are blessed with.

This investment seems almost too good to be true, but I have so much faith in the person who recommended it that I'm going to put all my retirement money into it.

In the examples given above, faith-based decisions resulted in accidents, hardships, or catastrophes. But negative outcomes are not always the case. For example, I might buy on faith a stock that goes up exponentially, thereby resolving any need for me to be concerned about retirement funding. (I may say I'm being rewarded for my faith, but my friends—old and new—could also rightly call me lucky.)

A faith claim is a claim of knowledge. Do not confuse this with hope. Hope is a wish, a desire. While there are those who maintain that believing in a thing strongly or intently enough can make it occur, wishing won't make something so. If a thing does not happen, it is not because someone didn't believe strongly enough.

Others think that advocating something more often, longer, or more loudly will make it true. Frequency, duration, and volume of a claim simply are not factors which make something true or false. A fervent, compelling, deep, sincere, firm, or strong faith is no better than simple faith.

Science (noun): *knowledge gained by evidence via observation, experiment, and logic*

There is an alternative way to know of reality, the opposite of faith. It is science, a method of accumulating knowledge through examining evidence.

While most of us don't consider ourselves scientists, we informally use the method of science in our daily lives, often without being aware we are doing so. We are experimenting all the time, but we go through these experiences without thinking of them as science. A toddler eating a first ice cream cone is doing an experiment. A second grader mixing paint colors is doing an experiment. A teenager learning how to ride a skateboard is doing an experiment. A chef adding a new ingredient to a recipe is doing an experiment.

> *Science is a way of trying not to fool yourself.* (Richard P. Feynman)

Let me illustrate the method of science with an example from a memorable road trip to Montana that outlines the formal methods by which we acquire scientific knowledge.

Step 1: Observe something in the natural world.

Note a fact, event, situation, etc.

After driving routinely on I-90 in Montana for a few hours, I took the exit for Bozeman. I slowed down as we entered a zone where the roadway was being ground up and prepped for resurfacing. There were bumps and ridges along all the lanes.

All of a sudden we heard and felt a terrible rhythmic thumping from the vicinity of the transmission or the wheels. When I slowed, it slowed. I put the car in neutral and coasted with the engine off. The alarming noise continued, so I suspected that a tire had gotten cut or torn or that we were riding a rim. I pulled over and checked the tires but could not see or feel anything amiss. This was serious. We couldn't continue like this.

Step 2: Ask a question.

What caused this? Does it affect anything else? Does it fit or improve the model, the theory?

My question: Do I have a tire problem?

Step 3: Has the question already been answered?

Typically scientists are aware of others who might be interested in answering the same question and where these researchers would likely publish their results. Even though, as Professor Frank H. Westheimer quipped, *A coupla months in the laboratory can save a coupla hours in the library*, it pays to check to see if an answer has already been found.

As I stood there, I noticed a big tire store across the road. Someone there must know tires and they could sell me a new one if one was needed. I drove slowly to the store, trying not to damage anything further.

Step 4: Identify a testable hypothesis.

This is an assumption, an answer to the question that must be *falsifiable*. That is, it must have the capability of being proved false so you can discard it if necessary. There may be many other possible answers.

I walked in and declared my hypothesis to the friendly lady at the counter: *I believe I have a tire problem.*

Step 5: Design an experiment.

This will test the hypothesis and/or deductions or predictions following from it—and either support it or disprove it. A scientist would also want to identify independent and dependent variables. What causes what? Establishing causality is often a significant task for the scientist.

I described my recent experience to the clerk and asked if a knowledgeable person could examine my tires. She responded that not all apparent tire problems are real tire problems. She could have someone look at the tires, but she suggested I do a simple no-cost experiment myself. A crossroad past the tire store had been recently repaved and was very smooth and straight. She suggested I drive a mile down that road, turn around, and come back. If the tires sounded normal, I could continue on my way. If the terrible noise persisted, I should stop in and they would look at it. She was quite confident she would not see me again. She saw many drivers each day with the same complaint. They called it *The Bozeman Chop*.

Step 6: Do the experiment.

Carefully record the experimental data. Others may wish to see if they get the same data or the data might later also help answer other related questions.

I took the test drive and soon discovered she was right. The car drove as smoothly as ever. At different speeds everything still felt normal. My hypothesis had been wrong.

Step 7: Analyze the results.

Research only supports or disproves an hypothesis. It does not *prove* anything. In science, uncertainty is never eliminated. If the results are not definitive to a pre-determined level of confidence, the hypothesis may need reformulation, the experiment may need to be redesigned, or the hypothesis may need to be rejected. Scientists learn much from failed experiments. Before the Wright brothers succeeded, many early *pilots* tested their hypotheses of how flight should work.

I was wrong; I did not have a tire problem. To my satisfaction I had disproved my hypothesis. The result was still a valuable and pleasant one, since it meant I could continue on my trip. The honest clerk's quick help in the tire store was much appreciated. The problem now appeared to be in the roadway. Other conditions could have been responsible, but they were not likely.

Step 8: Make results available to others.

Scientists understand, even expect, that others will examine their work and offer critical review. They may conceive a better experiment or alternative hypotheses or discover a delusion. Or they may agree it was well done and add the results to their current knowledge of the world and move on to another question.

In this instance, unfortunately, I couldn't easily inform other drivers what I had discovered. Plus, I reasoned, the question would disappear as the roadway got repaved.

The scientific method is designed to eliminate dishonesty and bias in order to uncover the truths of the natural world. It's not often this linear and simple. There could be many iterations at a single step, and while some steps could be quick and simple, other steps could be tedious and take years.

So, what's the difference between facts, laws, and theories? That is a part of science that many find confusing.

Scientific Fact:

Observed data, repeatable or verifiable observation. A scientific fact may answer a question, but it does not explain anything.

Scientific Law:

If an interesting or significant pattern in the data, the facts, is repeatable and always observed by anyone who is qualified and cares to look, it becomes a law. Some laws may be insignificant while others may be very significant. In our recorded history we have accumulated a wide variety of laws. Among the more famous in physical science are Boyle's Law, Hubble's Law, Ohm's Law and the Law of Conservation of Angular Momentum. Outside of science, among the more popular *laws* are Murphy's Law (*If something can go wrong, it will*) and the Law of Tumbling Toast (*Falling toast always lands jelly-side down*).

Scientific Theory:

A theory is the proposed conceptual framework explaining how or why the law works. This theory supports a model for understanding, a model that expands on previous models and allows us to predict other new results, predictions which can then be tested. Some important scientific theories are: Cell Theory, Atomic Theory, Chaos Theory, the Germ Theory of Disease, and Einstein's famous Theory of General Relativity.

Facts, laws, and theories can be and are regularly modified or even disproved by new experiments, observations, more accurate measurements (the greatest contributor), better understanding, and unexpected discoveries. Scientists are not at all concerned when laws and theories are replaced or updated with newer ones. The goal of science is to expand the knowledge base. It's a continuing adventure. This is what makes it exciting. As biologist and associate professor PZ Myers explains, *Science isn't in the business of proving anything, ever. It's a process for developing knowledge about the world, and we don't talk about proofs, because knowledge is provisional and changes as we learn more.*

Not all apparent bodies of knowledge are valid science. An apparent body of knowledge can be biased by politics, opinions, emotions, economics, time, irrational models, commercial interests, dishonesty, delusions, or measurement errors and is typically called *junk science.* Samples of popular but pure hokum pretending to somehow relate to a science-based reality range from astrology, Big Foot, and crop circles to voodoo, witchcraft, and zombies. Our ongoing task as scientists, as humans, is to sort the valid from the junk and use the junk only for entertainment or toss it.

Non-scientists may refer to ideas or hunches as theories, but a scientist calls those conjectures. To contribute to a theory, a conjecture must be formulated as a hypothesis and then tested. Either it fits and adds to the current model of reality or it is discarded.

It's not always that easy, of course. When you believe something based on evidence and the evidence changes, you have a reason to change your belief. When your belief is based on faith, when you *know* something without evidence, you have no reason to change your mind and doing so can be very difficult.

A law can be recognized for centuries even as the theories trying to explain it come and go.

Take gravity, for instance. It is part of our collective view of reality, part of our knowledge. We know that when an object is dropped it falls toward the center of the earth. Scientists have established that galaxies, stars, comets, moons, planets, people, paper clips, and rocks—in fact, all objects in the universe—are attracted to each other. Isaac Newton even came up with a mathematical formula, the Universal Law of Gravitation, that shows how the amount of attractive force between any two objects is related to their masses and the distance between them.

But how does gravity actually work? What is the origin of the attraction? There have been many theories. Someone long ago conjectured that all material things are in love and simply want to be nearer each other. Others have explained how gravity works by saying *God made it that way.* Neither theory is helpful in advancing scientific knowledge. Scientific theories are always and only checked against reality.

Sometimes the theory that is simpler becomes the preferred one. However, if a theory can also better explain related peripheral phenomena, it can dominate. That was the case with Einstein's Theory of General Relativity and its concept of curved space and time, which predicted and quantified previously unknown gravitational phenomena. GPS systems use this theory today to improve positioning accuracy.

Progress in science is very dynamic. Scientists continually discover new phenomena, formulate new laws based on them, and develop new theories. An old theory could be cast aside and forgotten. Or it could become part of a newer theory. In the 1960s, for instance, Alfred Wegener's 1915 theory of continental drift was refined when scientists were able to determine that drift was not limited to continents. Huge land- and water-covered sections of the earth's crust, called plates, were shifting. This led to the theory of plate tectonics.

There have been many notable conflicts between faith and science in history, some of which continue to this day. Let's look at two of the biggest.

Galileo

The process of science tends to eventually filter out errors. This can take centuries, however, especially when a new theory conflicts with a widely believed, strongly held earlier theory. Or when a new theory based on science conflicts with a prevailing theory based on faith.

Consider the theories of the shape of the earth and of the earth's relationship to the sun. While these eventually changed, there were those who, well into the 16th century, thought the earth was flat. (There are even some who believe this today, for that matter, and for the same reason—this verse from Matthew: *Again, the devil taketh him up into an exceeding high mountain, and sheweth him all the kingdoms of the world, and the glory of them.* (Matthew 4:8)) Explorer Ferdinand Magellan disagreed: *The church says the earth is flat, but I know that it is round, for I have seen the shadow on the moon, and I have more faith in a shadow than in the church.*

One of the earliest stories about where the sun went at night was that when the sun was out of sight each day, the Sun God was fighting a battle with an evil god. If you didn't worship the Sun God and sacrifice enough he might lose and the sun might not rise in the morning. And then you would really be in trouble. Forever.

During the 16th century the Catholic Church dominated in Italy and most of Europe, both religiously and intellectually, and it was quite adamant in claiming the sun rose in the east every morning because the sun, along with the stars and planets on celestial spheres, revolved around the earth once each day. That was their theory. They called it *sunrise*, a term we still use today, rather than *earthspin*. It was as safe and old as Plato, Aristotle, and Ptolemy and supported by the Bible as well: *He set the earth on its foundations; it can never be moved.* (Psalm 104:5) (See also: Psalm 93:1, Psalm 96:10, 1 Chronicles 16:30, and Isaiah 14:7) Furthermore, it just *made sense.* Scientists knew by then that the earth was round and all known objects were attracted to the center of it. Earth therefore had to be the center of the universe.

In 1609, German astronomer Johannes Kepler began to challenge this theory. Using data collected by Tycho Brahe, he published *Astronomia Nova,* in which he stated that the planet Mars followed an elliptical orbit around the sun with the sun at a focus of the ellipse. Today this path is called the First Law of Planetary Motion. This was more specific than Galileo Galilei's eventual theory, but it didn't seem to raise any red flags among Catholics, at least not in Germany.

Galileo experienced a different reaction in Italy when he published *Dialogue on the Two Chief World Systems*, a 1632 work benefiting from the existence of the newly discovered telescope and the teachings of Nicolaus Copernicus, a Polish Catholic priest. It promoted another theory about sunrise: that it occurred because the sun was fixed in the center of the solar system and the earth rotated on its axis once every day as it revolved around the sun once each year. This theory could better explain the phases of Venus, retrograde planet motion, and the orbits of moons around Jupiter. It could also describe planetary orbits while still accounting for the sun appearing in the east every morning.

Galileo's *theory* didn't change the facts, or the *law,* regarding sunrise. The observable pattern of the sunrise data remained the same. Whether he was right or dead wrong, the sun continued to appear on schedule in the east every morning.

His theory made the earth just another one of the planets, however. The universe did not revolve around the earth. It didn't even revolve around the sun. Man had lost his special place in God's creation. This was a significant and threatening concept for the Catholic Church. For centuries the church had taught that Earth and man were special and central in the scheme of God's creation. Cardinal Robert Bellarmine, writing to a Galileo supporter, Carmelite provincial Paolo Foscarini, in a letter dated April 12, 1615, said, in part:

I say that, as you know, the Council [of Trent] prohibits expounding the Scriptures contrary to the common agreement of the holy Fathers. And if Your Reverence would read not only the Fathers but also the commentaries of modern writers on Genesis, Psalms, Ecclesiastes and Josue, you would find that all agree in explaining literally (ad litteram) that the sun is in the heavens and moves swiftly around the earth, and that the earth is far from the heavens and stands immobile in the center of the universe. Now consider whether in all prudence the Church could encourage giving to Scripture a sense contrary to the holy Fathers and all the Latin and Greek commentators. . . . It would be just as heretical to deny that Abraham had two sons and Jacob twelve, as it would be to deny the virgin birth of Christ, for both are declared by the Holy Ghost through the mouths of the prophets and apostles. (*Modern History Sourcebook*, Fordham University: The Jesuit University of New York, January 1999)

Actually Galileo's theory said nothing of scripture nor Catholic theology. He was just following the evidence. It was the church that picked the fight.

The Catholic Church placed Galileo under house arrest in 1633 because he wouldn't abandon his heliocentric theory. At trial, the Inquisition of the Catholic Church ruled that he was *vehemently suspect of heresy*. His new book was banned. It was added to the Index of Forbidden Books and not removed until 1835.

Galileo was fortunate to be a well-known and likable Catholic. At least he wasn't subjected to St. Thomas Aquinas' sentence for heretics: *If forgers and malefactors are put to death by the secular power, there is much more reason for excommunicating and even putting to death one convicted of heresy.*

Over time, of course, Galileo's theory prevailed. In discussing it in 1992, over three-and-a-half centuries later, Pope John Paul II said, *The majority of theologians did not recognize the formal distinction between Sacred Scripture and its interpretation, and this led them unduly to transpose into the realm of the doctrine of the* ***faith*** *a question which in* *fact pertained to* ***scientific*** *investigation.* [Bold emphasis added.] (*L'Osservatore Romano*, November 4, 1992)

The pope then went further, issuing an apology, which lifted the Edict of Inquisition against Galileo and admitting that the Catholic theory was wrong and Galileo's was correct. He said again in part: . . . *The error of the theologians of the time, when they maintained the centrality of the Earth, was to think that our understanding of the physical world's structure was, in some way, imposed by the literal sense of Sacred Scripture . . .*

Science has continued to progress dramatically without much hindrance by errant faith-based theory. Today NASA's Kepler Mission, which has already identified more than 1,000 exoplanets (planets outside our solar system), is showing that our circumstances are not unique. In our own Milky Way galaxy there are over 100 billion stars. And there are over 100 billion galaxies. In fact, evidence is gathering that each star in the universe may have its own planetary system and many of these may have a planet suitable for life as we know it on our earth. The rush is on to design instruments that could allow the scientists to identify the chemical elements and molecules present on some of these planets and thus determine if they might support a familiar form of life.

Evolution

The evolution of living things is at the center of a modern example of faith and science in serious conflict.

> *And God created great whales, and every living creature that moveth,*
> *which the waters brought forth abundantly, after their kind,*
> *and every winged fowl*
> *after his kind: and God saw that it was good.* (Genesis 1:21)

Virtually every society in history has its unique story of how its world came to be.

Creationists believe and teach that God created all the plants, animals, and humans six thousand or so years ago and that these species have not changed much since. Their theory is supported by the Bible: *So God created man in his own image, in the image of God created he him; male and female created he them.* (Genesis 1:27, Mark 10:6)

At the same time, the Bible also states that living things can change or be influenced to change. Later in Genesis, it offers this example of genetic modification and artificial selection: *Then he placed the peeled branches in all the watering troughs, so that they would be directly in front of the flocks when they came to drink. When the flocks were in heat and came to drink, they mated in front of the branches. And they bore young that were streaked or speckled or spotted.* (Genesis 30:38–39)

> *What can be asserted without evidence can be dismissed without evidence.* (Christopher Hitchens)

Scientists have another theory about how species change, one based on observable and testable evidence.

Evolution, i.e., changes in inheritable traits over successive generations of biological populations, was already an observed and accepted phenomena in 1859 when Charles Darwin, a Cambridge-trained theology major, began publishing *On the Origin of Species by Means of Natural Selection, or the Preservation of Favored Races in the Struggle for Life*. Notice that he was writing about the **origin** of species. How do new species originate? Where do they come from? By what mechanism do they arise? How does it work? He stated his theory in his title: *by Means of Natural Selection. Favored races,* those that succeed in the competition for resources, he held, are able to reproduce and thus are preserved, a process that came to be called the *survival of the fittest.*

According to the Theory of Evolution, new creatures don't just pop up with a whole new set of characteristics. New characteristics appear slowly over many generations in a population. Eventually the population may transition to a new species, or it may split, with each branch possibly evolving into a new species. Since evolution is gradual and ongoing over very many generations, every creature that ever lived is a transition creature and of the same species as its parents. All species share common ancestry. All living creatures are *cousins*, i.e., related.

In each case, however, there is no first *Adam* creature. If you could align all the *Australopithecus australopiths, Australopithecus africanus, Homo rudolfensis, Homo ergaster/erectus,* and *Homo sapiens* (us) on a timeline, there would be no well-defined boundaries, no first of each species, no first human.

It is easy to see why Biblical literalists would find all this challenging.

> *The theory of evolution is no mere nuisance—it represents one of the greatest challenges to Christian faith and faithfulness in our times.*
> (Dr. R. Albert Mohler, Jr., 2011)

Creationist theory is simple. God did it. The Bible tells us so. (Recall how taking the Bible literally got the Catholic Church in a fix with Galileo?)

Adherents of creationism tried to get public schools in the US to include this theory in their curriculums. When it became clear, however, that it was unconstitutional for schools to accommodate pronouncements about God and the Bible in their science classrooms, proponents came up with *intelligent design* (ID), the theory that an intelligent designer created all the plants and animals on Earth.

This, too, was met with legal challenges. In 2005 the use of ID as an alternative to evolution in a science class in Dover, Pa., resulted in a famous court case: Tammy Kitzmiller, et al. v. Dover Area School District, et al. *A six-week trial over the issue yielded 'overwhelming evidence' establishing that intelligent design 'is a religious view, a mere re-labeling of creationism, and not a scientific theory,' said [U.S. District Judge John] Jones, a Republican and a churchgoer appointed to the federal bench three years ago.* (*Associated Press*, December 20, 2005)

Creationists say the earth is six to ten thousand years old.

Scientists say it is close to five *billion* years old.

Creationists calculate the age of Earth by relying literally on the Bible.

What have scientists used to determine the age of the earth? Evidence abounds. Here are just a few examples showing the earth is way over 10,000 years old.

The Vostok Ice Core, extracted in 2003 at Vostok, the Russian station in East Antarctica, is an ice core 2,083 meters in length. It documents—in layers similar to tree rings—the annual temperatures along with the air, dust, and CO_2 contents for each of about 400,000 years.

Other later cores by European and US scientists have independently corroborated the Russian ice core data and extended their reach to about 700,000 years ago.

Fine, well-preserved annual stone sediment layers in the Green River Formation in the Green River of Colorado, Utah, and Wyoming contain data covering 6 million years.

Radiometric dating, a technique developed more than a century ago that uses radioactive decay rates to date natural and man-made materials, indicates the earth is more than 4 billion years old.

As for where we came from in the first place . . .

Can a simple creature evolve into a more complicated one? Research confirms that it can. Evolutionary changes typically occur over many thousands to millions of years. Recently, however, two researchers at the University of Minnesota, William Ratcliff and Michael Travisano, using simple artificial selection, observed as 73 generations of single-celled brewer's yeast evolved into multicellular organisms. This significant transition occurred in just a few weeks.

Scientists today estimate that most of the creatures that ever existed on Earth have gone extinct. They also estimate that there are currently ten million species of plants and animals here. *The Encyclopedia of Life* (http://eol.org/) project is tracing the precise evolutionary relationships of more than a million of these species. DNA, first isolated in 1869 and modeled by Crick and Watson in 1953 as a double helix, is a molecule that encodes the genetic instructions used in the development and functioning of all known living organisms. It allows us to determine the paternity of a child or trace our ancestry back to possibly Attila the Hun. It also shows that humans evolved from animals, that we and chimpanzees, our closest living wild animal relatives, have common ancestors that lived about 6 million years ago. Over 98 percent of our DNA is the same as that of a chimpanzee. (More than 50 percent of it is the same as that of a banana too, dating back to when plant and animal life diverged approximately 1.5 billion years ago.)

Many processes account for altered DNA among living creatures over time, including symbiosis, mutations, sexual selection, differential reproductive success, epigenetic modifications, gene transfer, genetic hitchhiking, hybridization, viral insertions, copying errors, deletions, translocations, and genetic drift (a major factor). Some of these apparently random mutations result in creatures becoming better adapted to their surroundings. Sometimes it is the environment that changes, a circumstance that may confer an advantage to one creature over another. Mutations can also result in survival disadvantages, such as a susceptibility to conditions like breast cancer or color blindness.

We are witness to evolution and its consequences all around us. Flu viruses are evolving as fast or faster than we can develop new anti-flu inoculations. There's been a dramatic increase in drug-resistant tuberculosis and in staph and gonorrhea infections that are nearly untreatable. Super-weeds and -worms in cornfields seem to be adapting to the chemicals applied to kill them. This reminds us that evolution continues today and is not something that only happened long ago.

> *A Christian can accept the theory of evolution as a helpful explanatory model, provided he does not fall into the heresy of evolutionism, which views man as the random product of biological processes. . . .*
> (YOUCAT 42)

Still, there are many who continue to deny evolution, among them the Catholic Church, which included the statement above in its 2011 Youth Catechism of the Catholic Church. The Catholic Church pretends to accept Darwin's theory of how evolution works, albeit as a process guided by God, even as it embraces the idea of God injecting souls into humans beginning with Adam and Eve.

By the time of Darwin's death in 1882, he was accepted by the Anglican Church in England, which fittingly gave him a full state funeral and burial in Westminster Abbey near Sir Isaac Newton. In 2008 the Church of England apologized to Darwin for early on *misunderstanding* him.

But denying evolution is like denying gravity. It will continue no matter what you think. Evolution of living things happens. It is observed. It is well documented. This is so well established that we would be justified in declaring what we observe as the **Law** of Evolution or simply evolution.

In our science-based reality, there is the phenomenon of evolution and there is a theory of how evolution works. Even if Darwin's Theory of Evolution was dead wrong, while we searched for a new theory of how it occurs we would still be left with the fact that living things evolve. So, if you want to become world-famous overnight all you would have to do is develop a better theory of how evolution works.

Evolution is random variation followed by non-random selection.
(Richard Dawkins)

Evolution is what living organisms do. In fact, evolution can be found in any situation where variation followed by reproduction followed by selection occurs repeatedly over a period of time. This applies not only to living creatures, but also to folk tales, urban myths, languages, superstitions, political party planks, religions, cultural memes, the childhood game of telephone, and even sports.

Take baseball, for instance. If professional baseball did not already exist, it would be impossible to get 20,000 people to pay 25 dollars each to watch a group of young men taking turns hitting a ball with a bat and running around a ninety-foot square back to where they started. Likewise, without a paying audience it would be impossible year after year to pay a group of eighteen uniformed players 3 million dollars each to play the game. The audience and players need each other. Neither can continue to exist without the other.

So the question is: Which came first, the large-paying audience or the multimillion-dollar players? The answer, of course, is that they evolved together—and continue to do so. They evolved so slowly over the years that it's safe to claim there was no first baseball game. (Note: There was no Intelligent Designer involved.) Interestingly, the bat-n-ball game evolved into a big business, a self-sustaining enterprise.

We should also note that baseball players and fans cooperate in the venture: The players support the fans and the fans support the players. Evolution is now understood to not only be about competition, but also about cooperation.

Which came first, the hen or the rooster? The simple obvious answer is this: They evolved together. Regarding the famous chicken-or-egg question, it appears the egg came before the chicken since evidence shows that reptiles preceded chickens and laid eggs. Evolution also brought the chicken and chicken-egg together.

If you wish to reject evolution, you also need to reject large blocks of supporting well-accepted science and history. In physics this includes: radiometric dating (verified and calibrated via dendrochronology, the scientific method of dating based on the analysis of patterns of tree rings), isotope decay, and energy functions. In paleontology: fossil layers and fossil records (transitional forms and dinosaur feathers). In geology: stone layers and age of layers. In comparative anatomy: embryology and homologous traits (human tails and dolphin legs). In genetics: synteny mapping and phylogenetic systematics, which reconstructs evolutionary history.

There are non-scientific ways to argue for the soundness of evolution as well.

The Bible says that *And God saw every thing that he had made, and, behold, it was very good.* (Genesis 1:31) But we can point to many *mistakes* in nature, many ways things could have been better created if the creator (or Intelligent Designer) was at least as smart as we are. Like omitting the human appendix (at least its problems) and making our jaws larger to accommodate wisdom teeth (better yet, skip the wisdom teeth altogether). Why create a foreskin and then decree that it be cut off at the first opportunity? And whose idea were viruses anyway? And mosquitoes, gallstones, the lancet liver fluke, malignant melanoma, and poison ivy?

God is presumed to be omnibenevolent, as well as omnipotent, omniscient, and omnipresent. But the idea of an omnibenevolent being creating an imperfect universe filled with much painful and deadly competition for territory and food is akin to the notion of a loving, caring parent telling her toddlers to go play in the busy street.

Could an omnibenevolent God create the smallpox virus that killed over a half billion innocent people? If He did and He declared, *it was very good,* how does He feel about the fact that a crew of humans (another of his creations) came along, declared it not good, and destroyed it, eliminating it from the planet? How does He feel about the work being done by the World Health Organization, the US Centers for Disease Control and Prevention, UNICEF, Rotary International, the Bill & Melinda Gates Foundation, and others to add the polio virus to the list of creatures destined for intentional extinction later in this decade?

How does He feel, for that matter, about the living things humans have created through artificial selection—Holstein cows, Black Lab dogs, Ragdoll cats, Russet Burbank potatoes, Lady's Blush roses, Rhode Island Red chickens, and the popular seedless grape? The seedless grape is so modified (yes, a GMO) that it can't even reproduce itself by itself. How does He feel about those we have modified genetically, including one of my favorites, the transgenic (part human, part goat) animals which produce recombinant human antithrombin in their milk, used as an anticoagulant for the prevention of clots before, during, and after surgery?

Finally, how does He feel about humans creating for human use, new living creatures with *no* ancestors? In 2010, Craig Venter and Hamilton Smith announced the first synthetic life form, a bacterium with new human-designed chemically synthesized DNA, an artificial genome whose immediate ancestor was a computer instead of another bacterium. And more recently a group of college students has created an entirely synthetic fully functioning yeast chromosome. In 2015 we are learning that the polio virus DNA has been modified to allow it to contribute to the elimination of certain brain tumors. We are also seeing significant activity in research related to the regeneration of organs and limbs.

[*Newsflash* (January 2016)—Josiah Zayner began selling a home kit that allows you to use the recently developed CRISPR gene editing tool to create new creatures or edit genetic sequences, on your kitchen table. Of course, this has possible dangers (think old horror movies). Some liken it to assembling an atomic bomb on your kitchen table.]

As we enter a grand age of evolutionary biological science, the mystery and miracle surrounding the creation of creatures is disappearing. Humans are discovering another way to edit humanity.

Reason

Reason is the process of using logic to uncover new truths. It can be used by Christians or atheists, by conservatives or liberals. Scientists use it in order to expand what is known beyond what is directly observed, to know the unobservable: the past, the future, that which is immeasurably small or large. It can help them understand how to harness nature to work to our advantage as well as to understand the consequences of messing with nature.

Arguments are the foundation of reasoning. They use **if/then** statements like these:

If *I spend all my money today*, **then** *I won't have any money tomorrow*.

If I *drive at this average speed*, **then** I *will arrive at your house in three hours*.

Note the form of these statements: **If** *premise*, **then** *conclusion*. In an argument, a valid inference is the logical process to get from a premise to a conclusion. The inference could be valid and if the premise is false the conclusion may be false.

Note, too, that it's possible to think of ways—by using an invalid inference, for instance—that the conclusions reached above could be wrong. Someone could spend all her money one day, for instance, and get paid the next day. Someone could drive at a set speed and arrive at someone's house later than predicted because he took the wrong road.

That's where formal logic comes in. It is the study of how to get from premises to true conclusions. Formal logic has developed over thousands of years and has arrived at many laws of valid inference grouped as deductive, inductive, statistical, syllogistic, propositional, etc.

All the various formal logic branches agree on the following important logical property:

If the premise is true with a valid inference, then the conclusion must also be true.

If a system of logic cannot guarantee this property, it is worthless.

> *Indeed, to argue according to rigorous rational criteria is to guarantee that the results attained are universally valid.* (John Paul II, FIDES ET RATIO, 1998)

We all use reason regularly, often without being aware that we're doing so. A common use is in games. There's no sure-fire way to win in Tic-Tac-Toe, but by putting your first X in a corner you can increase your chances. And by making rational moves, you can always prevent your opponent from winning and thus achieve a tie. In Hexapawn, played on a three-by-three grid similar to that in used tic-tac-toe, there are no ties. Someone always wins. In order to be that player, however, you must allow your opponent to move first.

Reason is also used to diagnose problems that cannot be directly observed. Your doctor hears your description of your symptoms, asks some questions, and possibly orders some tests, which are essentially scientific experiments to help test her hypothesis. Depending on their results, she makes a diagnosis using observation and reason and prescribes a treatment. More than two thousand years ago Eratosthenes, employing similar thought processes, used observed measurements of the sun to deduce the spherical shape of the earth and calculate its size—all without going more than 500 miles from home.

You use rather basic science and reasoning when you are driving along and observe that all the cars around you are driving at no more than the posted speed limit—an unusual occurrence. You theorize that there must be a highway patrol officer somewhere nearby, because an accident up ahead would almost surely bring traffic to a stop. You start looking around and soon spot a patrol car parked in the median.

Your theory of how patrol cars affect traffic has allowed you to predict something you had not yet observed. This theory and conclusion was adequate for the situation, even though you didn't rule out all other possible causes for the reasonable rate at which people were driving. If future events or conclusions depended on it, however, you would need to identify, examine, and test all those other possible causes.

Our observations and reasoning are sometimes imperfect. That's what makes magic so popular: our surprise and delight at being shown that reality is not as we thought. We almost always respond, *How did you do that?* We just know that there must be a rational explanation.

The same thing happens in science. Author and biochemist Isaac Asimov once noted that *The most exciting phrase to hear in science, the one that heralds new discoveries, is not 'Eureka!' but 'That's funny . . .* When something unexpected happens during an experiment, we want to find out why. How did it do that? Why did it do that? We know there must be a rational explanation.

To find those explanations, scientists regularly tell others what they have observed and then reasoned. They want to know: Did I observe that correctly? Did I discover something new, or did I simply make a mistake? Is that a validly reasoned conclusion? That is because oftentimes what appears to be a valid rational argument, but due to an incorrect inference, can actually be a fallacious or invalid one.

How does that happen?

One common logical fallacy is to **Argue from Ignorance**:

I do not understand what could have existed before the Big Bang, therefore God must have done it.

I don't see how a bird wing could have evolved, therefore it must not have.

Another logical fallacy is the **Ad Hominem Argument**, criticizing the person making the statement rather than the substance of it:

You claim that you are innocent of a crime, but you are a felon and thus I say you can't be trusted. Your claim of innocence must be false.

Another is the **Appeal to Belief** fallacy:

My friends all believe that you are guilty, therefore it must be true that you are guilty.

My medical school professors and my textbooks tell me that gastric ulcers are caused by excessive stress, therefore it must be true.

Circular Reasoning is another common problem.

I *know God exists because it says so in the Bible, and I trust the Bible because it was inspired by God.*

The final guide to the interpretation of scripture is scripture itself. (Answers in Genesis)

Also common is the **Two Wrongs Make a Right** fallacy:

Punishing someone by executing them is okay because it is only used on people who have killed someone.

Adults often use the **Argument from Authority** to make opinion/testimony sound like fact:

I'm in charge, therefore I'm right.

Another common fallacy is **Proof by Example(s)**:

Catholicism is false because of what some priests did to children.

Islam is false because of what some Muslims did to the World Trade Center towers.

The Biblical story of Babel is true because we speak different languages.

Then there is this trio of really common mistakes in reasoning.

Correlation Equals Causation:

Roosters crow in the morning, therefore they cause the sun to rise.

The number of visible umbrellas correlates to the amount of rain falling, therefore more umbrellas will make it rain more.

Absence of Evidence is Evidence of Absence:

I see no evidence that my house is on fire; that proves it is not on fire.

I haven't found any evidence that God exists, therefore God does not exist.

Confirmation Bias (believing something, then proving it with some plausible-sounding statement):

Belief: *Trickle-down economics is true.* Proof: *A rising tide raises all boats. If the rich get richer, everybody will prosper, right?*

Last but not least, **Tradition** *(this is the way it's always been)* and **Wishful Thinking** are also not valid arguments. And repeating an invalid argument over and over doesn't make it a valid one.

Formal logic is designed to avoid these pitfalls. Here are some of the principles it adheres to in order to achieve valid reasoning:

If the premise is true with a valid inference, then the conclusion must also be true.

Two contradictory statements cannot both be true at the same time. While this idea may seem obvious, it is often ignored.

A false premise in a valid logical argument always leads to a

conclusion that may be true or false. This is a common defect in theological arguments. (Remember—we are looking for a conclusion we can count on being true.)

Example

Statement A: **If** *this bird egg drops five feet to a concrete floor (true premise),*

then *the bird egg will break* (conclusion—true).

Statement B: **If** *this bird egg drops five feet to a concrete floor* (which is a false premise since it only dropped five inches onto a foam pad), **then** *the bird egg will break* (conclusion—may be true or false).

If a statement is true, the statement's converse may be true or false.

Example

Statement: **If** *the flashlight battery is dead*, **then** *the flashlight bulb is dark* (true).

Converse: **If** *the flashlight bulb is dark*, **then** *the flashlight battery is dead* (true or false, since the bulb may be burnt out or the switch is defective).

Some theologians suggest that a variation of this called **Inference to the Best Explanation** can be useful in acquiring knowledge of reality. It goes something like this:

If fact **B** is observed and if situation **A** always causes **B** to occur, **then** it is fair to conclude **A** likely exists and **A** is the cause of **B**.

If there are multiple **A's** that can plausibly cause **B**, **then** the *best* or *most likely* of these **A's** exists and is arguably the cause of **B**.

Perhaps some examples might be helpful to understand this better.

Example 1.

B: The flashlight is dark (observable fact).

A1: The battery is dead. (The most likely cause, according to the experience of those considering the problem.)

A2: The bulb is burnt out.

A3: The switch is defective.

Conclusion: A1 is perceived as the most likely explanation and therefore is declared as the cause of **B**. Install a new battery, which may or may not solve your issue.

Example 2.

B: The patient has stomach ulcers (observable fact).

A1: Excessive stress/worrying increases production of stomach acid, say doctors, which causes ulcers.

A2: A specific kind of bacteria (*H. pylori*) is present in the stomach.

A3: The patient gets too little exercise.

Conclusion: A1 (stress and worry) causes **B**. To get better, stop worrying. (This was the accepted diagnosis and treatment until a few decades ago. It is now recognized that most stomach ulcers are caused by *H. pylori*.)

Example 3.

B: Our universe originated with a Big Bang (known, rationally deducible fact).

A1: An omnipotent God did it, according to theologians.

A2: Quantum physics says particles can pop into existence from nothing.

A3: A black hole exploded again.

A4. First there was a Big Collapse.

Conclusion: A1 caused **B**. In other words, God exists and caused the Big Bang.

One can easily see that if A1 is the preferred answer or is judged by those in authority to be most likely, it still might not exist or it might be dead wrong. We cannot depend on this attempt at logic. It is working logic backwards. This is another example of *confirmation bias*. The folks who use it call it *abductive inference*. You may call it simply guessing.

To summarize:

False premises and invalid arguments lead to useless conclusions.

A true premise used in a valid logical argument always leads to a conclusion that is true, and this conclusion can then be the true premise in the next valid logical argument, etc. This procedure is common in mathematics and science.

Science has not always been as influential as it has been in the last few centuries. Religious faith and theocratic governments ruled from the Dark Ages through the Middle Ages. Then, in the 17th century, science, using mathematics as a tool, began to be able to accurately predict the future and measure the previously unmeasurable. Libraries are filled with books describing feats of reasoning that led to new, surprising, and dramatic conclusions. During the Age of Enlightenment, also called The Age of Reason, rational thought, together with free thought, free speech, and free press, replaced theism and tradition. Natural Law replaced *the Divine Right of Kings* in government and science. Democracies and republics replaced kingdoms and papal states. Science found its place as the best means of assessing truth and reality. Progress quickly followed.

William Herschel discovered the planet Uranus in 1781. But its mathematically derived orbit didn't always fit the observed path. In fact, the discrepancy got wider over time. One proposed explanation was that it was being acted on by the gravitational pull of another, unknown planet. Working independently, Urbain Jean Joseph Le Verrier and John Couch Adams calculated the location of a planet that could be responsible for the discrepancy. On September 23, 1846, the planet was found at the mathematically predicted location. It became known as Neptune, after the Roman god of the sea. [Newsflash (January 2016)—Astronomers announced the mathematical discovery of another planet. They are calling it Planet Nine or Planet X. It has not been spotted yet.]

Helium is the second most abundant element (after hydrogen) in the universe. In 1868, using the newly developed technique of spectroscopy, scientists discovered helium on the sun. It took another 28 years before it was found on Earth. Interestingly, this same technique has recently detected methane on Mars, which likely originates from biological (living) processes.

In 1892, Hendrik Lorentz argued mathematically for the existence of electrons. Five years later, J. J. Thomson experimentally demonstrated their existence. Today these subatomic particles flow through the computer circuits of your cellphone, processing billions of app instructions per second, enabling you to connect with nearly anyone in the world.

The idea of bodies so massive that even light could not escape from them was first put forward by John Michell in a letter written in 1783 to Henry Cavendish of the Royal Society. In the 1960s theoretical work showed that black holes, long considered a mathematical curiosity, were generic predictions of general relativity. While they cannot be visually identified, today the presence of black holes can be inferred through their interaction with other nearby matter. Physicists are now at work trying to verify hypothesized gravity waves and dark matter.

Mathematics is a logic force magnifier: It gives you access to all its accumulated, proven, logical thinking. When mathematicians supply the formula for the area of a circle or the derivative of a function, you only see the *tip of the iceberg*, the results of logical arguments, writings, and thoughts that went into developing and proving those formulas. **Mathematical reasoning** is a structure of symbols and logic that offers a systematic shorthand way of reasoning and quantifying. This can be applied to all kinds of situations, including predicting everything from where the waterline should be painted on a ship before the ship is launched or the future value of a stream of payments, to determining the maximum number of spheres that can fit into a box.

One way to reach a valid conclusion using mathematical reasoning is to start by assuming (using it as a premise) the *opposite* of what you intend to prove. In **Indirect Proof**, also commonly called **Proof by Contradiction**, if an assumption leads to a contradiction or an absurdity, then the opposite of this contrary assumption must be true. For example: If you want to prove x = z, you begin by assuming that x does not equal z. If this leads to a contradiction, then the opposite of this assumption must be true: x *is* equal to z.

Suppose, for example, that your partner opens a fresh deck of cards. You ask her, with no sleight of hand, to shuffle it and pick a card, look at it and without showing the card to you, turn it face down on the table. You ask her to tell you which card is turned down. She says it is the ten of diamonds, but she often lies. Your challenge is to prove the facedown card is the ten of diamonds and she is telling the truth without touching, moving, or looking at it.

Here's where we use logic. The facedown card either is or is not the ten of diamonds. We want to prove that it is the ten of diamonds, so we'll assume the opposite, that it is *not* the ten of diamonds. Logically, that means the ten of diamonds must be still in the deck.

You go through the deck, one card at a time, looking for that ten of diamonds. Not finding it is a contradiction of your statement that the ten of diamonds must be still in the deck. Your original assumption that the facedown card is not the ten of diamonds must be wrong. Therefore it must be the ten of diamonds. She was telling the truth.

Mathematical results also can seem to fly in the face of common wisdom.

Say you're playing a coin flipping game. You win a point if you call the correct result before the flip. You just witnessed a string of seven heads. Should you call heads or tails on the next flip?

In reality, it doesn't matter. While we expect in the long run that heads and tails will land about the same number of times, on any given flip the probability of heads is still 50 percent. The coin cannot remember what happened on previous flips. Frequently, however, a player will choose tails on the next flip, thinking that tails must be *due*. This error in logic is so common that it has been given a name: The Gambler's Fallacy.

Likewise, what would you say if I asked you to purchase a lottery ticket with the numbers 2, 3, 4, 5, 6, and 7? Probably something along the lines of: *Don't be silly. That would be very unlikely to win. It probably has never happened in the history of lotteries.*

But, in fact, that ticket has exactly the same probability of winning as any other set of six numbers you might choose.

Daily Dilemmas Between Faith and Science

Faith and science, as two different ways of knowing something, can sometimes converge on the same reality. Either can be correct or not. What if they do not agree? What if they agree? Let's consider this example, in which we are driving toward an uncontrolled intersection, which sometimes has cross-traffic that requires us to stop and yield:

First, let's assume we drive based on faith.

- If faith (some drivers don't look) says there is cross-traffic and I stop, but scientific evidence says there is no cross-traffic, then I have stopped unnecessarily, wasted gas and time, and possibly slowed traffic.
- If faith says there is no cross-traffic and I proceed while the scientific evidence says there is, I may cause an accident.

Next, let's assume we drive based on science.

- If science says there is cross-traffic and I stop, but faith says there is no cross-traffic, then I'm still doing the right thing.
- If science says there is no cross-traffic and I proceed even though faith says there is cross-traffic, then I'm still doing the right thing. No accident occurs.

If faith and science agree, it doesn't matter if we drive by faith or science.

If faith and science disagree, and we drive based on faith, the result can be inconvenience or a possible accident.

If faith and science disagree and we drive based on science, the result is *doing the right thing*, independent of faith.

It follows that in situations where faith and science disagree,

science rules. It doesn't matter that one believes an event will unfold another way. Reality has a way of prevailing.

Every accident is a dramatic instance of a conflict between faith and science.

When you decide to look at your cell phone instead of the road ahead, you are basing your actions on your faith that nothing dangerous or threatening is occurring or will occur in front of you. You choose to believe the road is not slippery. You choose to believe you can stop faster than the driver in the car in front of you. You choose to believe the other driver knows what he is doing. When a crash occurs, you may call it an accident, but you chose where to look and how to act based on faith rather than assuming other drivers might do something stupid. Faith told you that you were safe. Then reality intervened.

There are many who accept faith-based certainty as genuine, trustworthy knowledge. Some claim that faith can enhance or support science or that science can support faith. In truth, there are no in-between or compromise relationships between science and faith. They are opposites, like a win and a loss or an open safe and a locked one.

Human beings sometimes hold views that are contradictory. Not every thought a scientist has is necessarily scientific. And not every thought that a theologian has adheres only to the tenets of faith. It is possible to keep the natural and supernatural realms separate in one's mind though cognitive dissonance can result. Case in point, the work of the BioLogos Foundation, a community of Christians *committed to exploring and celebrating the compatibility of evolutionary creation and biblical faith . . . for the good of the church*. In truth, its exploration appears to be an effort to adjust faith to what science has discovered or find ways to rationalize away the contradictions. The *celebrating the compatibility* part seems particularly premature and misplaced since:

Faith accepts without proof.
Science rejects without proof.

Faith stops curiosity.
Science encourages curiosity.

Faith is supported by authority.
Science is supported by evidence.

Faith claims cannot be tested.
Science claims must be tested.

Faith is seeing with your eyes closed.
Science is seeing with your eyes open.

Faith forbids skepticism.
Science encourages skepticism.

Faith allows you to believe.
Science allows you to know.

Faith supplies certainty.
Science endlessly pursues certainty.

Faith does not need facts.
Science depends on facts.

Faith handicaps and undermines education.
Science supports education.

Faith embraces miracles.
Science denies miracles.

Faith rewards thoughts from the heart.
Science rewards thoughts from the mind.

In faith, mystery is the answer.
In science, mystery is the challenge.

Faith values belief at the cost of ignoring evidence.
Science values evidence at the cost of changing belief.

After a coin toss, I declare, without looking at the coin (i.e., faith), *Heads*. After looking at it (i.e., science), you declare, *Tails*. We cannot both be correct. We can't both be even partially right. Either faith and science agree or they do not. And when they do not, science and its evidence trump faith.

Tolerance is a noble virtue in helping us get along with one another, but it is not a virtue when the question is whether *creationism* should be included in the science curriculum at a secular or nonsectarian school. Science-based and faith-based theories are not equivalent. A faith-based theory that the earth is about six thousand years old cannot demand equal standing with the science-based theory that the earth is nearly five *billion* years old. They cannot command the same credibility.

Those who are offended by the conclusion of a valid, logical argument with a true premise deserve no sympathy. Reality is not influenced by one's attitude towards it. It simply is what it is. Being happy, confident, or satisfied with an invalid argument simply because it is not upsetting is not only short-sighted, it can also be dangerous. Remember President Ronald Reagan's famous, but paradoxical quote, *Trust but verify?* He was admitting that with regard to the Soviet Union's nuclear weapons program, trust (faith) was nice, but insufficient.

Reality is that which, when you stop believing in it, doesn't go away.
(Philip K. Dick)

In the real world, evidence and rational thought predominate.

If your monthly bank statement arrives with a major discrepancy in it along with some apparently valid data, you investigate until you have a complete and rational explanation. You don't apologize for insisting on that. You question and audit until the numbers balance.

At a murder scene an investigator needs to construct a scientific theory of what happened, who did it, and possibly why. The detective makes observations, conjectures, and forms testable hypotheses and tests them. Anything that doesn't fit is discarded or investigated further. The investigation continues until the theory of how things happened fits to the greatest degree possible and is supported by evidence.

Someone is arrested and charged with the crime.

At this point the case goes to trial. The prosecutor explains why the police investigator's theory should be accepted, and the defense attorney argues that the theory is irrational or unsupported by the evidence and should be rejected. A judge or jury determines the validity of the theory—or tries to, at least.

Unfortunately, sometimes the evidence is flawed or incomplete. Humans are not always perfectly rational and completely honest. The work of the Innocence Project has led to the exoneration of more than 300 people wrongfully convicted of crimes (usually murder) in which DNA evidence is available to be tested or retested. The vast majority of those freed were convicted on the basis of mistaken eyewitness testimony; over half of the original convictions resulted from faulty forensic evidence.

The goal of our legal system is rational, evidence-based justice. As these exonerations demonstrate, we are still trying to get it right. (Of all the cases taken on by the Innocence Project, by the way, about 43 percent of clients were proven innocent, 42 percent were confirmed guilty, and evidence was inconclusive in 15 percent of cases. In about 40 percent of all DNA exoneration cases, law enforcement officials identified the actual perpetrator based on the DNA test results that led to an exoneration.)

When I was a child, I spoke and thought and reasoned as a child. But when I grew up, I put away childish things. (1 Corinthians 13:11)

Among children in Western countries some of the biggest mysteries in their lives center on Santa Claus.

Where did all these toys come from?
Why don't we ever see Santa?
How does he get into our house?
How does a big man like Santa get down a small chimney?
What happens if you don't have a chimney?
How does he know what everyone wants?
How does he know if you are sleeping or awake?
How does he know if you are good or bad?

If we waited until children were 18 before presenting them with the story of Santa Claus, it would quickly be rejected with a retort on the order of *Are you kidding?* Three- and four-year-olds might question how he knows whether they've been good or bad, but they have no difficulty perceiving the trimmings—the Santa suit, white beard, and husky, jolly voice—as *real.*

As they learn how to reason, the questions become more difficult. *How can Santa get everywhere in the world in just one night? How can a sleigh carry that much weight? Can reindeer really fly?* The mysteries we invent and post-hoc rationalizations we make in an attempt to answer these queries are endless.

After I heard two classmates in my first grade class on the playground say that there was no Santa, I asked my dad about it. I cried when he confirmed what they'd said. I had believed. I had been fooled. The revelation shocked me at first, but then all the pieces came together. I had wondered about the chimney thing and flying reindeer. It never seriously occurred to me, though, that Santa did not actually exist.

On the one hand, the mysteries and miracles of Santa Claus had just vanished. On the other hand, Dad explained that even though I knew the truth about Santa I would still get toys and other presents. I agreed to help keep the fable going so my younger sisters and brother wouldn't find out too soon.

I had slipped through one of those virtual barriers that separate childhood from adulthood. The transition made me realize that in order to believe in Santa I had had to accept a lot of other incredible (i.e., unbelievable) phenomena. I began to wonder if this was true of other so-called mysteries. More and more as I went through life, whenever additional miracles were needed in order to explain the original mystery, warning horns beeped, red lights flashed, and orange flags waved. My skepticism grew. As time went on, I also learned that William Tell, King Arthur, and Lady Godiva did not exist.

Skepticism is the immediate antidote for faith, but science is the long-term cure. I found the new reality refreshing: the universe was rational and discoverable.

When seeking entertainment, variety, or escape we enjoy the irrational, fictional, artistic, miraculous, magical, and mysterious. But when our money, life, health, or freedom is involved, we usually rely on rationality and evidence. We use it to put our shoes on the correct feet. When we cross the road, we don't let our fate depend on faith; we look both ways. When we're up to bat, we don't swing based on faith; in that split-second as the ball approaches, we decide, based on evidence, if we will swing or wait. And when it comes to treaties to eliminate nuclear and conventional ground-launched ballistic and cruise missiles with intermediate ranges, we *trust but verify.*

In matters large and small, faith takes a back seat to scientific evidence. By depriving themselves of a way to understand the world around them, science deniers handicap themselves severely. Simply put, science, powered by evidence, reason and its logic, works.

In Part II we will consider the second definition of faith: a set of religious beliefs, usually relating to the supernatural.

Part II

Religion

Question with boldness even the existence of God . . .
(Thomas Jefferson)

Religious beliefs usually center on worship of a deity or deities, supreme beings called gods. These beliefs are wide-ranging, from creation stories to the characteristics of a supreme being and a supernatural world beyond human experience. The supernatural world has all manner of interesting characters, including ghosts, angels, goblins, dragons, and zombies, but we will give our attention here to how religions explain how the supernatural world affects the natural world and causes it to be the way it is.

Religious beliefs try to answer existential questions such as these: *Where did I come from? Why am I here? How should I behave?* And probably most importantly: *What happens to me after my death? How can I enter the supernatural world? Can I influence the supernatural world to affect the natural world in my favor?*

The mysterious, invisible gods of the supernatural world leave no physical evidence as they ostensibly wield amazing powers to influence our world. Still, religions make truth claims about the supernatural.

Faith (noun): *1. knowing something with no evidence; 2. a set of religious beliefs, usually relating to the supernatural*

We examined the first definition at length in Part I. Now let's dig in to the second definition: a set of religious beliefs, usually relating to the supernatural.

In religion, as in politics and sports, we find comfort in joining with others of similar belief. Religious beliefs and associated rituals, along with those who follow them, tend to coalesce into religious groups or religions. Also called *faith traditions, religious traditions*, or *religious heritages*, religions each have their own sets of core beliefs, creeds, and *professions of faith*. These are often compiled in a *catechism*, a series of fixed precepts, often presented in the form of questions and answers that are used for instruction. Some beliefs are shared by many religions while others are unique. A belief within any one religion can be absent or totally contradicted in another. Buying into a religion is usually like buying an annuity. It offers a payback later—how to live, how to live forever, how to be *saved*. Or as Mark Tooley of the Institute on Religion and Democracy noted in 2014, *All religion, even its most permissive forms, aims on some level to tell its adherents how to live. Otherwise it has no purpose.*

Both the definition of faith that was introduced in Part I and the definition introduced above are combined in this statement from the Catholic Catechism:

Faith is the theological virtue by which we believe in God and believe all that he has said and revealed to us, and that Holy Church proposes for our belief, because [He] is truth itself. (CCC 1814)

And both definitions of faith are used here:

. . . faith is seen mostly as a set of doctrinal mores, whereas it is primarily a free act by which one entrusts oneself to God. (Vatican City, VIS 8 October 2014, Fourth General Congregation)

Religions generally have some form of organizational structure, congregate regularly in places of worship (their church or temple or mosque), and claim a superior moral authority. Each religion holds that theirs is the true religion, the true faith. Each religion claims that its origin stories and miracle stories are uniquely true. (Applying the rules of logic noted in Part I, we should notice here: If one religion is right, the others can't be right as well. By the same token, if one religion is wrong, all can be wrong.)

The most common yet most hidden temptation is our lack of faith. (CCC 2732)

Members come to their religions by birth, baptism, confirmation, or conversion. There is little tolerance among believers for any criticism of their religion. This is nothing new: Religious intolerance has existed from biblical times. In the Inquisition of 1492, Jews were forced to convert to Christianity or leave Spain. In the 20th century, they were consigned to concentration camps for their faith.

Things are no better in the 21st century. Some believers are so fervent in their religious faith, in fact, that their groups must be separated by walls. In Northern Ireland *peace lines,* border barriers built in the 1960s to minimize inter-communal violence between Catholics and Protestants, still divide neighborhoods today: a study released in 2012 indicated that 69 percent of residents believe that the walls are still necessary because of potential violence. In 2008, Christians in many states in India were told to convert to Hinduism or flee. In 2014 Islamist insurgents in northern Iraq issued an ultimatum to the Christians there to convert to Islam, pay a religious levy, or face death by the sword. For other examples, study the Sunni/Shiites in the Middle East, especially the Syrian *civil* war, or the conflict between Christians and Muslims in Nigeria, the most populous country in Africa. In Saudi Arabia, atheism is viewed as a crime equivalent to terrorism. These conflicts are not only international phenomena, either. The FBI reported that in 2011 about 20 percent of hate crimes in the US were motivated by religious bias.

Most civil wars today, says journalist Christiane Amanpour, *are still over religion.* The only way to settle which religion is correct seems to be war. Think of the war between Islam, Christianity, and Judaism as *the regionals.* The bigger war, between faith and science, will be *the finals,* the championship—winner takes all.

Wouldn't it be great if religions with their contradictory faiths spent more time debating each other rather than killing each other? But debates would draw attention to the history of how and why religions differ and allow attendees to make note of contradictory claims. Participants would be reminded that their faiths have evolved from other faiths and were established when earlier members decided another faith was wrong and *voted with their feet.* The three main Abrahamic religions, for instance, have evolved from a common ancestor. Christianity, Islam, and Judaism all accept that God revealed himself to Abraham, who was willing to kill his son to prove his faith. (Couldn't he climb a tree, move a boulder, eat fire, or do something else to prove his faith?) These religions share other common traits, including monotheism and recognizing Adam as the first man and God as the source of moral law. In other ways, they diverge greatly.

There are well over ten thousand (possibly as many as a million) identifiable religions in recorded history, many of them offshoots of other religious traditions. Each religion has had its corresponding *God* or god(s), from Allah, Christ, Helios, Horus, and Itzamna to Maya, Mithra, Nut, Odin, Pan, Ra, Thor, Typos, Venus, Yahweh, and Zeus.

This has led some to advance the idea that humans might have an in-built God belief, something that they contend lends more legitimacy to faith in the supernatural. This might seem plausible until one realizes that nearly every human society in history also believed in monsters of various kinds or fire-breathing dragons, the best known creature that never was. Through this lens, the question *Who is your God?* has equal standing with *Who is your monster?*

In ancient Egypt, each pharaoh favored particular gods. Three millennia BCE, royalty needed pyramids (with their contents) to reach the afterlife and immortality. In the third century BCE, the cat goddess Bastet and other animal deities were still being worshipped. The ancient polytheistic Greeks had many famous gods—from Aphrodite, Apollo, Ares, Artemis, and Athena to Poseidon and Zeus. As recently as the 1940s, Japan's emperor was regarded by some as a god.

Throughout history various gods have been credited for lightning, famine, death, storms, seizures, sunshine, disease, wind, and rain. As modern science allowed us to understand these things better, the corresponding gods and religions receded into the background. Some, but not all, have vanished. Even today we see segments of the population attributing to God phenomena that science has already fully explained without a God premise.

While interesting religions of recorded history are simply too numerous to list beyond a few such as Zen, Druze, Mithraism, Sikhism, and Wicca, we can't ignore some famous recent *startups*. About 20 new religions or religious sects appear each year in the United States alone. While some eventually wither (or self-destruct), others are growing in numbers. Here are just a few and the names of those who helped start them.

Branch Davidians (David Koresh)
Church of Google/Googlism (Matt MacPherson)
Church of Scientology (L. Ron Hubbard)
Fundamentalist Church of Jesus Christ of Latter-Day Saints (Warren Jeffs)
Heaven's Gate (Marshall Herff Applewhite, Jr.)
Lakewood Church (Joel Osteen)
The People's Temple of the Disciples of Christ (Jim Jones)
Saddleback Church (Rick Warren)
Westboro Baptist Church (Fred Phelps)
World Changers Church International (Creflo Dollar)

Religions start as small, stigmatized cults, almost always founded by men and with a male God. Only a few survive for more than a few decades. Some religions evolve over time and survive, but most have simply vanished along with their gods.

Key to the success of a religion, of course, is having more people joining the religion than leaving it. Some of this is accomplished through conversion. Evangelists and missionaries plead with individuals to abandon their religious traditions and adopt new faiths. While some choose to follow a new faith based on the values espoused by the religion, others decide depending on which missionaries and their supporting religions seem most credible, successful, and worth following. Or have better music, more entertaining services, or more convenient hours of worship.

At the same time, religions need to hold on to those whom they recruit or those born into the faith who indicate doubts or concerns. This explains why many religions advocate punishing apostates. They have done this through banishment (Quaker), disconnection (Scientology), shunning (Jehovah's Witnesses), isolating (Mormonism), or execution (Islam). Exiting from Catholicism apparently incurs a form of excommunication, according to recently updated Catholic Canon law regarding *opting out.* (See CCL 751, 1364)

Major World Religions

According to the *World Fact Book*, in a world population of more than 7 billion people, 33.39 percent are Christian, 22.74 percent are Muslim, 13.8 percent are Hindu, 11.7 percent are nonreligious or atheist, 6.77 percent are Buddhist, 0.35 percent are Sikh, 0.22 percent are Jewish, and 0.11 percent are Baha'i. Other religions account for another 10.95 percent. Some faiths are nearly as old as recorded history; on others the ink in their holy books is barely dry.

Hinduism—One of the oldest major world religions is also, with some 1 billion members, one of the largest. Hinduism is *poly*theistic: It venerates many gods. Early holy books include the Vedas and the Puranas, which date back to two to three millennia before the current era, i.e., 3000–2000 BCE. Vaishnavism, a sect of Hinduism with 350 million members, began around 300 CE. Shaivism, which has 200 million members, began in 150 BCE.

Judaism—Begun about 2000 BCE, Judaism is *mono*theistic; it venerates one god, known as Yahweh. The holy book of Judaism is the **Torah**, from which parts of the Old Testament of the Christian Bible is derived. There are some 14 million members, including 4 million adhering to Reform Judaism, begun in 1810 CE, 2 million adhering to Orthodox Judaism, begun in 1851 CE, and 4.5 million adhering to Conservative Judaism, begun in 1886 CE.

Buddhism—Dating back to about 600 BCE, Buddhism is a *non*theistic religion: It has no god. (Jainism and Confucianism are also *non*theistic religions.) The beliefs of Buddhism's 500 million members are shaped by teachings (http://en.wikipedia.org/wiki/Original_Teachings_of_the_Buddha) attributed to Siddhartha Gautama, commonly known as the Buddha. A major precept is the concept of reincarnation, the cycle of birth, death, and rebirth which continues until an individual reaches *nirvana*—freedom from karma, extinction of desire, passion, illusion, and the empirical self, and attainment of rest, truth, and unchanging being.

Christianity—Christianity is a monotheistic religion that worships God in three persons, the Father, Son (Christ, after whom the religion is named), and the Holy Spirit. In its various forms, the faith has some 2.2 billion members. Roman Catholicism, with 1.1 billion members, began in 0–300 CE. It is led by a pope. Eastern Orthodoxy, with 400 million members, dates back to 100–400 CE. Protestantism, which has some 700 million members, began in the 1500s CE. The holy book of Christianity is known as the **Bible**. Adopted as official in Constantinople in 325 CE, it has two parts, the **Old Testament**, which consists of the sacred writings of the Jewish people, and the **New Testament**, which is ascribed to God-inspired disciples.

Islam—Islam originated in 610 CE on Mount Hira, near Mecca, where the angel Gabriel delivered the first of many revelations, the **Qur'an**, from God to Muhammad. Known as Muslims, its members worship a single god, Allah. Around 700 CE the religion split into two branches, Sunni (1 billion members) and Shi'a (150 million members).

Mormonism—This offshoot of Christianity began in 1820. The vast majority of its 13 million adherents belong to the Church of Jesus Christ of Latter-day Saints, which teaches that the angel Moroni appeared to Joseph Smith in upstate New York and told him the location of two golden plates. Before these disappeared, their contents were translated by Smith to English and became the book of Mormon.

Jehovah's Witnesses—This was begun in 1931 CE, as a restorationist Christian denomination with nontrinitarian beliefs. Today it has some 7 million members.

Scientology—This body of beliefs and related practices was created by science fiction writer L. Ron Hubbard, beginning in 1952, as an expansion of his earlier system, Dianetics. It teaches that people are immortal beings who have forgotten their true nature. It has about half a million members. In 1994 it was recognized as a religion by the IRS, which granted tax exemptions to its entities in the US. In 2013 the Supreme Court in England declared it an official religion there as well.

By global standards, any of these dominant world religions is a minority religion. Catholics, for example, are matched in number by the Islamic Sunni, and worldwide at least four times as many people believe something other than what Catholics religiously believe. If you are religious that means that no matter which faith you profess, more people of the world have different beliefs and, in many matters of faith and the supernatural, they think you are wrong.

That wouldn't be a big deal if not for the fact that many religions claim to be the one true faith. If you really are wrong, then you are likely destined for eternal damnation. This alone is enough to justify studying and analyzing religions.

Seventeenth-century French philosopher Blaise Pascal used this concept to make a deceptively simple case for maintaining one's faith. In *Pascal's Wager* he argued that given the possibility that God actually does exist and assuming that belief or unbelief in God leads to spending an eternity in heaven or hell, a rational person should live as though God and heaven exist. If you believe, pray to God, go to church, behave, give generously to the collection, die, and there is no

God, no harm done, except for missing out on a few illicit pleasures. If you're wrong, you're just finished. That's the end. On the other hand, if you die as an atheist and God *does* exist, being wrong packs a very different punch: You get eternal damnation for not believing.

There are problems with choosing to believe as a form of insurance. Pascal's Wager was centered on Christianity, but everyone thinks his or hers is the one true god. If the safe bet is to believe in God, which one out of thousands deserves my faith? What happens if I choose the wrong god? Allah, say. My bad. Or if my parents or grandparents chose the wrong god? God knows I tried, right? Praying/adhering to a faith *just in case* is as rational as placing garlic in your doorway *just in case* there are vampires.

Also, if I believe, just to be safe, won't God know I am faking it? Isn't He omniscient? Can He be fooled?

What if you're seduced by this argument into trying to believe in a god, just to be safe, and the evidence for belief is just not there? It's like trying to believe in Santa again. Can you look at evidence that it is daytime and try to believe it is nighttime? Can you make yourself believe the earth is flat? Or that it is less than 10,000 years old? Or that fire-breathing monsters lurking around the corner will get you? Will your beliefs be based on evidence or threats? (Keep in mind that if your beliefs are based on evidence, you don't have to choose. There's only one option.)

It is generally accepted, at least publicly, that everyone is entitled to his or her own religious beliefs. Things get sticky, though, when one religion has opinions about the beliefs in another.

Christians believe that Jesus is God, was born of a virgin (something he has in common with Horus, Perseus, Buddha, Mercury, Krishna, Mithra, and Romulus and Remus, all of whom are credited with miraculous births), that he died on the cross and after a few days was resurrected by a different incarnation of himself and later ascended into heaven.

Judaism holds that Jesus is a false Messiah, noting that according to the Jerusalem Talmud, *If a man claims to be God, he is a liar.* (Ta'anit 2:1)

In Islam, Jesus didn't die on the cross and is only a messenger of God, though he can perform miracles.

Neither Islam nor Judaism believe in the resurrection of Christ.

Islam claims all children are born Muslim.

Catholics claim all children are born in Adam's original sin.

Religions also commonly assert that things happen for a reason, for God's purpose. In the case of calamities like earthquakes, plane crashes, and deadly childhood diseases, those trying to express sympathy to survivors typically offer one of two possibilities:

- The victims were evil or, though totally innocent, were co-existing with those who were evil (think members of the Westboro Baptist Church picketing the funerals of soldiers killed in Iraq or Afghanistan to protest gay rights or televangelist Pat Robertson claiming that Hurricane Katrina was God's punishment for America's abortion policy and thus the 1,800 who died deserved their fate).
  This doesn't seem rational, especially in the case of young victims.
- God had a plan and had to make these people suffer in order to accomplish a greater good.
  This second option doesn't make sense either since we assume God can accomplish anything with as little as a poof! Why not just do the greater good and skip the disaster with its collateral damage? After all, God is omnipotent—*For God nothing is impossible.* (Luke 1:37) He is also ethical and omnibenevolent—*One may never do evil so that good may result from it*, and *The end does not justify the means.* (CCC 1789 and CCC 1753)

The Catholic Catechism addresses this contentious issue this way:

The fact that God permits physical and even moral evil is a mystery that God illuminates by his Son Jesus Christ who died and rose to vanquish evil. Faith gives us the certainty that God would not permit an evil if he did not cause a good to come from that very evil, by ways that we shall fully know only in eternal life. (CCC 324)

But . . . Doesn't this virtually scream, *THE END JUSTIFIES THE MEANS?*

Martyrdom, dying for your faith, is a characteristic of a number of religions, including Christianity and Islam. On June 29, 2013, Pope Francis said, *Let's learn to lose our lives for Christ, like a gift or a sacrifice. With Christ we lose nothing.* When Pope Benedict XVI visited the United Kingdom in 2010, Muslim demonstrators, referring to the Crusades, reminded him that justice and sharia law were on the way.

Holy warriors of various sorts are, indeed, trying to topple secular governments all over the world, but not only Muslims. While among Muslims, Shiites are seen as God's warriors, Catholics, when confirmed, are said to have become *soldiers of Christ.*

Muslims and Christians each want to convert the other even if they have to die doing it. (And isn't it interesting that under certain conditions when a Muslim kills for his religion he gets 72 virgins in his afterlife, but when a young Catholic merely touches himself the wrong way and doesn't confess it to a priest, he could be condemned to eternal damnation?) And each faith thinks its motives and methods are superior to the other's. Convicted-Watergate-conspirator-turned-prison-preacher Charles Colson epitomized that state of affairs when he said, *Christians will give their lives and die for what they believe, and all through the years they have done so, but Islamists are very different. We will die for what we believe. They will kill for what they believe.*

Sometimes religions turn to secular governments for protection, lobbying them to assist in defending their faiths from criticism by declaring such criticism *blasphemy* and making it illegal.

Islamists in Belgium and Iran have demanded that the European Union adopt anti-blasphemy laws. In support of this, Saudi King Abdullah bin Abdulaziz proclaimed, *I demand a UN resolution that condemns any country or group that insults religions and prophets.* This is interesting since the Qur'an insults Jews and Catholics: *The Jews say, 'Ezra is the son of Allah,' and the Christians say, 'The Messiah is the son of Allah.' That is their statement from their mouths; they imitate the saying of those who disbelieved [before them]. May Allah destroy them; how are they deluded?* (Qur'an verse 9.30, QuranX.com)

Catholics in Mumbai supported a long prison term, under Section 295(a) of the Indian Penal Code, for Sanal Edamaruku, an Indian author and rationalist who used the science of leaking water pipes and capillary action to explain a miraculous, weeping statue of Christ.

The United States government is officially religion neutral: *Congress shall make no law respecting an establishment of religion, or prohibiting the free exercise thereof. . . .* (Amendment 1, US Constitution) The very first item in the Bill of Rights mandates that government stay out of religion and religions stay out of government, and yet Christians routinely enlist the aid of federal, state, and local governments to support and enforce practices like this:

Christian rituals (prayers at government meetings)

icons (statues and crosses erected on public property)

mottos (*In God We Trust* on coins and *under God* in the Pledge of Allegiance)

official holidays related to Christianity (Christmas)

beliefs (the Ten Commandments posted on government property)

forbidden behaviors (anti-same-sex marriage amendments to state constitutions and the US Constitution, laws against birth control, laws forbidding the desecration of venerated religious objects, and *blue laws* restricting or banning business activities on Sunday)

school financing (government-financed vouchers for private religious schooling)

core education (anti-science *intelligent design* and *teach the controversy* legislation)

social order (laws against blasphemy)

privilege/endorsement (preferential tax treatments only since majority rules)

religious public relations (the National Day of Prayer)

conscience protections (faith-based opt-outs of legislative requirements)

The US government should be protecting the rights of people of all religions as well as those of no religion. That's what the founding fathers intended when they set up a secular government to protect us from Anglicanism, Christianity, Judaism, Islam, and all the other possible religions to come, like Mormonism and Scientology.

There is no mention at all of a deity in the Constitution, which came into force in 1789.

In 1797, the newly founded federal government proclaimed in Article 11 of the Treaty of Tripoli that the *Government of the United States of America is not, in any sense, founded on the Christian religion.* This treaty, submitted to them by President John Adams, was passed unanimously by the US Senate.

In 1819, James Madison, the key author of the Bill of Rights, credited *the total separation of the Church from the State with manifestly increasing the number, the industry, and the morality of the Priesthood and the devotion of the people.*

Some claim the United States is a Christian theocratic nation because on our coins it says *In God We Trust*. That phrase is a recent addition, historically speaking, having been placed there in the 1950s as part of a *Red Scare* anxiety.

There are those who would like to take that foundation and build upon it, however—to make the United States a true, full-fledged Christian theocracy. Theocracy is a form of government in which clergy have sovereignty over a territory and official policy is either governed by officials regarded as divinely guided or is pursuant to the doctrine of a particular religion or religious group. This is the form of government in Iran and Saudi Arabia. *Dominionists* think it should be the basis of government in the United States.

At a conference in 2005, popular American pastor D. James Kennedy described the scope such a government would have: *Our job is to reclaim America for Christ, whatever the cost. As the vice regents of God, we are to exercise godly dominion and influence over our neighborhoods, our schools, our government, our literature and arts, our sports arenas, our entertainment media, our news media, our scientific endeavors—in short, over every aspect and institution of human society.*

The Catechism of the Catholic Church contains related ideas: *By constantly evangelizing men, the Church works toward enabling them 'to infuse the Christian spirit into the mentality and mores, laws and structures of the communities in which [they] live.'* (CCC 2105)

And on the Anglican front: *The Church of England has complained that the promotion of fundamental values such as democracy, the rule of law, individual liberty and mutual respect and tolerance in schools could undermine Christian teaching.* (National Secular Society, November 14, 2014)

Then there are Iran and Saudi Arabia, Islamic theocracies ruled by religious sharia law. We are all too aware of how these regimes have reacted in recent years to human rights reforms and values commonly accepted elsewhere in the world and in other religions.

Anjem Choudhury, former head of *islam4uk,* stated in a British video press conference, *Nowadays, people worship idols which are more intellectual like democracy, liberalism, freedom and so on. So, these need to be destroyed as well and replaced with worshipping and obeying Allah.*

Hizb-ut-Tahrir, a political party whose ideology is Islam, says this about itself: *Hizb-ut-Tahrir is an ideological party that makes sovereignty exclusively to the [Sharia] ideology; it does not allow reality to be the source of the solution, but rather the subject of the solution . . . It does not alter nor distort the rules of Islam to suit reality. . . . work towards the return of the Islamic State as it was before, the leading superpower in the world.* (english.hizbuttahrir.org)

Religious Evolution

Despite such rhetoric, in modern times countries commonly have transitioned from theocratic to secular governments, rather than the other way around. International integration has also resulted in countries becoming more ethnically and religiously diverse. This has left many nations facing the challenge of not only dealing with the role of a particular religion in their governments, but also dealing with the role government should play in resolving conflicts between competing religions and between people who are religious and those who are of no religion.

In 360 CE Emperor Constantine built Hagia Sophia, the Eastern Orthodox Christian cathedral in Istanbul (Constantinople), Turkey. It was rebuilt in 537 CE by Emperor Justinian. In 1453 the Ottomans conquered the Constantinople church and Mehmet II ordered it be converted to an Islamic mosque. In 1935 the secular Republic of Turkey, at the behest of Ataturk, converted this to a museum where public prayer was prohibited. The religious-secular tug of war continues today with Turkish leaders in 2013 talking of converting the Hagia Sophia back to a mosque, and, of course, Christians are objecting.

In 2012 Norway amended its constitution to end a thousand-year tradition of supporting an official Christian religion. Spain, Italy, France, Germany, Finland, England, Denmark, Sweden, Canada, Japan, South Korea, and Australia are also good examples of the transition to the secular.

Gallup, Inc., a global polling company, reported in 2004 that continent-wide in Europe, just 15 percent of people attended religious services. This corresponds with a mathematical model published in 2011 by Abrams, Yaple, and Wiener that suggests religion is dying out in nine countries. They conclude: *The model indicates that in these societies the perceived utility of religious non-affiliation is greater than that of adhering to a religion, and therefore predicts continued growth of non-affiliation, tending toward the disappearance of religion.*

Or as George Leonard Carey, the former Archbishop of Canterbury, put it, more simply, in 2013: Christianity is just *a generation away from extinction.* In 2014, at the first meeting of the Archbishops' Evangelism Task Group, John Tucker Mugabi Sentamu, the 97th Archbishop of York, said, *Next to worship, witness is the primary and urgent task of the Church. Compared with evangelism everything else is like rearranging furniture when the house is on fire. Making disciples is at the heart of our Christian faith and our Anglican tradition.*

If you want to know the future of Christianity, says Pastor Rick Warren, *it is the developing world. It's Africa, it's Latin America, and it's Asia.* (Then again, with the Chinese buying African farmland, developing resources, and moving people there at an accelerating pace, Christianity could miss its chance on that continent in coming decades.)

The drift towards secularism in America has not been as extreme as that in Europe, but it has been well-documented, especially over the last half century.

In 2007, Gallup reported that 63 percent of Americans said that they were a member of a church or synagogue. In 1937, that number was 73 percent. In 1948, 91 percent of Americans identified with the Christian faith; in 2008, 77 percent of Americans did so.

Gallup surveys have also repeatedly asked this question: *Do you believe that religion can answer all or most of today's problems?* In 1957, 82 percent responded *yes*, but in 2009 just 57 percent did.

On the other hand, the number of people who are not affiliated with any faith tradition (*nones* in the parlance of the Pew Forum on Religion and Public Life) is growing. A 2007 survey Pew conducted put the number of *nones* at 15 percent. In 2012 that number was 20 percent. In 2015 the *nones* took another dramatic increase, rising to 25 percent of the US population. (The percentage may be higher even than that: It is reasonable to suppose that some people are reluctant to disclose that they follow no religion.)

Colleges also evolve from sectarian to secular institutions. A number of well-known schools that began as religious schools, usually to educate clergy, are now considered non-religious. (That is true at the University of Virginia, too, though when Thomas Jefferson founded the university, he declared that it should not have a divinity school and he banned the teaching of theology, stating that *The priests of different religious sects . . . dread the advance of science as witches do the approach of daylight.* Over the years, however, the university has added a religious studies department and a non-denominational chapel.)

Catholic institutions are changing as well. St. Louis University, a Jesuit institution, has dropped its requirement that *[t]he [p]resident shall be a member of the Society of Jesus*. The University of St. Thomas in St. Paul, Minnesota, not only changed its bylaws to allow a non-priest to become its president, but in 2013 it selected a laywoman to head the university.

It remains to be seen what impact this will have on what the Cardinal Newman Society has called *a crisis in Catholic higher education.* A study by Georgetown's Center for Applied Research in the Apostolate (CARA) in the United States shows that Catholic students who attend Catholic colleges and universities are more likely to move *away* from the faith than to have their faith strengthened. The study also found that 32 percent of Catholic students at Catholic schools attend Mass less frequently by the time they graduate, while only 7 percent attend more frequently. Roughly 8 percent of the graduates reported leaving the Catholic faith during their college years, while students who had entered the Catholic Church during their college years accounted for only 4 percent of the graduates. The survey found a strong trend away from accord with church teachings on issues such as abortion and same-sex unions. CARA also notes that the number of Catholic ordinations per year in the US has decreased from about 1000 in the 1960s to about 500 today.

Clearly, all this has caught the attention of the church hierarchy. In 2008, Pope Benedict XVI lectured Catholic colleges in America on the role they were to play:

The Church's primary mission of evangelization, in which educational institutions play a crucial role, is consonant with a nation's fundamental aspiration to develop a society truly worthy of the human person's dignity. At times, however, the value of the Church's contribution to the public forum is questioned. It is important therefore to recall that the truths of faith and of reason never contradict one another. . . . Teachers and administrators, whether in universities or schools, have the duty and privilege to ensure that students receive instruction in Catholic doctrine and practice. . . . I wish also to express a particular word of encouragement to both lay and religious teachers of catechesis who strive to ensure that young people become daily more appreciative of the gift of faith. (Meeting with Catholic Educators, April 17, 2008)

Pope Francis reiterated this in 2014, saying, *Each of the baptized, whatever their role in the church or the educational level of their faith, is an active agent of evangelization.* (Pope Francis, January 15, 2014)

The facts may tell you one thing. But God is not limited by the facts. Choose faith in spite of the facts. (Joel Osteen)

Religions are by nature authoritarian: God's laws are not subject to change, not open to influence by the laws of man. And yet religions change, they evolve. How?

Faith is . . . not subject to findings of science, contends Jan Paulson, past president of the General Conference of Seventh-day Adventists. In reality, however, successful religions continually adjust to science, albeit with a time lag—four hundred years in the case of Galileo, and about a hundred years in the case of women being allowed to lead a closing prayer at the semi-annual gathering of the Mormons, something which didn't occur until 2013.

Science, meanwhile, progresses independent of religion, something that enables it to be nationalistically neutral. China, Russia, France, the United States, and Iran all study the same DNA, nuclear, projectile, computer, and economic theories. Science never asks: What does this religion or that country think of this scientific theory? Religion may confront science, but within science, religious beliefs are not a factor, except sometimes as a distraction or temporary obstacle (e.g., stem cell research, birth control).

Only 18 percent of Americans say they know a scientist, whereas nearly everyone knows a preacher or two. Still, science does quite well in the competition with religion for supporters. Why does science lead?

In religion, faith is an asset. In science, faith is a liability.

Religions build on faith, fear, and mystery, while science builds on skepticism and evidence.

The natural world is filled with evidence; the supernatural world is an evidence void.

Religious faith uses mystery as an explanation; science uses mystery to identify fertile fields for study.

Religion claims to know the truth, while science only gets ever closer to the truth.

Religions teach that faith is a virtue (but that a person with a faith that disagrees with your faith is always wrong). In the world of science, it is accepted that a person with a theory that contradicts yours may be right.

Religion argues that its dogma is and always was correct; science constantly tries to disprove or improve its own theories. In theology if a contradiction is identified, it is explained away as something miraculous or: *God works in mysterious ways*. In science or mathematics if a flaw or contradiction in theory or conclusion is identified, it is studied until the contradiction is eliminated.

While science and its theories are only slightly concerned with blasphemy, religions want protection from or the punishment of anyone exhibiting irreverence toward things they consider sacred or hold in high regard. Religions claim to encourage tolerance and understanding among their adherents, but do not practice it with regard to others. Because religious dogma is seen to have originated in the divine realm, it does not invite compromise or questioning. Science invites challenges—which it answers with evidence.

Religious leaders would like you to think they are teaching you how to be a better, more moral person by teaching you the eternal values embedded in their religions. They maintain that moral leadership is top down.

Actually, it's just the opposite. When it comes to change, religion is most often *bottom up*. Change, when it happens, almost always originates at the lowest levels, with the people, the masses. Religions and preachers ultimately follow, kicking and resisting change all the way—because religions need adherents in order to survive.

Those in power at the apex of a faith's hierarchy rarely think it is a good idea for it to change or splinter. Like evolving animals after a population split, some religious entities quickly die off or drift off into oblivion while others thrive. For excellent examples, refer to the Protestant Reformation in the 16th century or the schism taking place today in the Anglican Communion and the Episcopal Church over the issue of homosexuals in the ministry, which has resulted in a civil lawsuit in Charleston, South Carolina, in which a bishop is asking the court to decide which of two competing bishops is the *legitimate* bishop.

Consider the following sample of changes that have occurred in religious and moral stances since biblical times.

Slavery—The Old Testament says this about slavery:

Your male and female slaves are to come from the nations around you; from them you may buy slaves. You may also buy some of the temporary residents living among you and members of their clans born in your country, and they will become your property. You can will them to your children as inherited property and can make them slaves for life, but you must not rule over your fellow Israelites ruthlessly. (Leviticus 25:44–46 (NIV))

Different nations woke up and outlawed slavery at different times. In the United States that happened in the mid-19th century during the Civil War. The Southern Baptist Convention was formed mainly to support slavery. It acknowledged that heritage in 1995 with this resolution:

Resolution on Racial Reconciliation on the 150th Anniversary of the Southern Baptist Convention

WHEREAS, Our relationship to African-Americans has been hindered from the beginning by the role that slavery played in the formation of the Southern Baptist Convention; and . . .

Be it further RESOLVED, That we ask forgiveness from our African-American brothers and sisters, acknowledging that our own healing is at stake; and . . .

Homosexuality—A number of faiths are still wrestling with this issue.

Individuals have recently pushed the Evangelical Lutheran Church in America (ELCA), Episcopal Church, Presbyterian Church, and the United Church of Christ to accept homosexuality. CNN's Dan Merica reports that Mormons are *softening [their] attitudes* toward the gay rights community. But innocent children of gays are still not accepted. Soon after the Boy Scouts of America(BSA) dropped their gay ban, the official Catholic Church's BSA liaison announced the new status is *not in conflict with Catholic teaching*. The Mormons and United Methodists are also accepting the new BSA position.

Pope Francis acknowledged pressure on this issue with this defensive-sounding statement: *If someone is gay and he searches for the Lord and has good will, who am I to judge that person?* In another statement he said, *For example, many think that changes and reforms can take place in a short time. I believe that we always need time to lay the foundations for real effective change*. More recently Cardinal Dolan said, *The pope's message was that we don't look at people first and foremost as their sexual orientation or sexual identification; we look at him (sic) as a child of God, made in his image and likeness.* (September, 2013) What makes this notable is that the previous pope had labeled homosexuality as *an intrinsic moral evil.*

Attitudes continue to evolve. *A majority of US Christians (54%) now say that homosexuality should be accepted, rather than discouraged, by society, including 70% of Catholics, 66% of mainline Protestants, 51% of members of the historically black Protestant tradition and 36% of evangelical Protestants and Mormons.* (Pew Research Center Survey, reported December 23, 2015)

Circumcision—In 2012 a German court ruled that this religious practice amounted to grievous bodily harm. Catholic, Protestant, and Muslim leaders voiced objections. The court had to choose between religious freedom and the well-being of a child. In 2013 Muslim and Jewish leaders petitioned the Parliamentary Assembly of the Council of Europe to rescind its anti-circumcision position.

Birth control—The Catholic Church officially forbids it, while over 82 percent of Catholic women in the US view it as morally acceptable.

Women's rights—Women have long advocated for voting rights and equality, and in the last two centuries a number of Christian religions have accepted women into the ministry. There is still no place for women in the hierarchy of the Catholic Church, however, nor in many other religions, for that matter.

As personal, cultural, and societal views have changed so, eventually, have precepts in the religions within those societies. When opposition by individuals and groups has become strong enough, religions have reassessed and modified their guidelines about a wide variety of practices—everything from eating with a fork and forbidding cremations, inoculations, and money lending to chattel slavery, human sacrifice, polygamy, the sale of indulgences, sterilization, suicide, and witchcraft.

The Catholic Church, however, is still holding onto the witchcraft of exorcism: *Thus devout prayer has such efficacy that it is a most powerful means of casting out demons; for there is a certain kind of demon which is not cast out but by prayer and fasting.* (See YOUCAT 273 and CCT)

In religions the evolution from denunciation to acceptance can be slow, taking decades and centuries, and be nearly invisible. For a religion, changing its stance on something is like you dropping your pants—embarrassing, if noticed. Though they fear these changes mean corruption, worldliness, or abandonment of the gospel, religions regularly reinvent themselves. Like all living creatures, they are simply adjusting to their surroundings, having no plan or goal other than to survive. Making things up as they go along, playing

Calvinball on a grand scale, they label these adjustments as *reinterpretations, a pastoral revolution, modernizing, doctrinal development, growing in communion, opening up broader horizons, interpreting today's reality, pushing the boundaries, shifting in practice, softening, pastoral discernment, church renewal, growth,* or they simply change focus from dogma to mercy. Dogma and traditions, standards of right and wrong, alter over time and space. Some adjustments result in survival advantages. Those that don't enable them to cope with changing surroundings lead to extinction.

In 1965, after about 1,500 years of Catholic anti-Jewish histrionics, Pope Paul VI admitted the Jews of today are not responsible for Jesus's death:

... *True, the Jewish authorities and those who followed their lead pressed for the death of Christ; still, what happened in His passion cannot be charged against all the Jews, without distinction, then alive, nor against the Jews of today.* (Nostra Aetate, Declaration on the Relation of the Church with Non-Christian Religions)

This gave Catholics official permission to be nice to Jews. More importantly, it signaled that people are not responsible for the behavior of their ancestors. We will revisit this in a later part in light of Catholicism's view of Adam, Eve, and the claimed consequences today of their sin of disobeying God and eating an apple.

In April 2007 another long-held Catholic concept, that of *limbo*, was retired. And in 2010 Pope Benedict XVI announced that in exceptional circumstances (in Africa) the use of condoms is acceptable. (What about all the poor souls that went to everlasting hell while condom use was declared always a mortal sin?)

The Catholic Church has no problem with heart pacemakers, artificial knees and hips, insulin shots, Viagra, and intraocular plastic lens replacements. However, it views both birth control and in-vitro fertilization (credited for the existence of more than five million humans) as evil/sinful, because they are not *natural.* (By the way, it is also not natural for humans to fly and yet we have found ways to do so; on average, more than 8 million people fly every day.)

More recently, Germany's Cardinal Reinhard Marx indicated support for divorced but remarried Catholics to receive communion after a suitable penitential period. (Currently one needs a Catholic annulment of the previous marriage before remarriage and reception of holy communion.)

To explore how to handle these and other issues, the church called an Extraordinary General Assembly of the Synod of Bishops on the Family in 2014. It was called *Extraordinary* since it demanded *immediate attention for the good of the entire Church* (USCCB, October 8, 2013) on not only divorce but also the effects on the family of war, immigration, domestic violence, polygamy, inter-religious marriages, and cohabitation. (This is striking, isn't it? If the Bible and official creed is divinely inspired, isn't it odd that the Catholic hierarchy is only inspired after the people are inspired? One would think that getting marching orders directly from God would put the church in a position to lead, never taking a false step, rather than follow.)

Times change and we must change with them. (Pope Francis, Homily, Santa Marta, October 23, 2015)

There is evidence of evolution in other faiths as well.

The Book of Mormon claimed blacks earned their black skin as a curse from God—until 1978, when God finally delivered to Mormons the divine inspiration that they could abandon racism and extend sacraments and priesthood responsibilities to blacks—as long as they are men. Women are not allowed to be priests. (If you find all this laughable, please don't be too noisy when you cackle: *Loud laughter* is still considered unholy and impure in Mormonism.)

The advance of medical procedures such as vaccinations, in-vitro fertilization (IVF), and blood transfusions have prompted changes in a number of faiths. While Jehovah's Witnesses still believe transfusions of whole blood are forbidden by God, official church position on this is evolving. Today members are taught that the use of blood fractions such as albumin, immunoglobulins, and hemophiliac preparations are *not absolutely prohibited* and a matter of personal choice.

Teachings about some subjects are sometimes contradictory within various religions and between them. Take the prohibitions listed in the Old Testament of the Bible, for instance.

According to Genesis 1:21, *God created the great sea monsters and every living creature that moves, of every kind, with which the waters swarm, and every winged bird of every kind. And God saw that it was good.* And yet in Leviticus 11:10 it says that eating shrimp is abhorrent.

In Exodus 20:13, Moses conveys the commandments God gave him on Mount Sinai, including the sixth commandment: *Thou shall not kill.* In Numbers 31:17–18, Moses tells his followers to kill every male among the little ones and every woman who has known man by lying with him, but to keep the virgins for themselves. And in Exodus 35:2, the people of Israel are enjoined to put to death anyone who works on the Sabbath.

Consider, too, what Catholicism and Islam say about masturbation.

While the Catechism of the Catholic Church states, *the Church does not demonize masturbation,* (YOUCAT 409) it also states that . . . *masturbation is an intrinsically and gravely disordered action . . . contrary to its purpose.* (CCC 2352) In the same section it also states, *To form an equitable judgment about the subjects' moral responsibility and to guide pastoral action, one must take into account the affective immaturity, force of acquired habit, conditions of anxiety or other psychological or social factors that lessen, if not even reduce to a minimum, moral culpability.* My question is this: How close to zero does *a minimum, moral culpability* get? This doesn't sound like the automatic mortal sin it was in my youth. Anyone who attended Catholic schools then would attest to the evolution that has occurred there.

Islamic law addresses the question of masturbation as well. Apparently Islam is okay with the practice—unless it occurs during the dawn-until-sunset fast observed during the month of Ramadan, whereupon the fast becomes void. (This leads me to wonder if there might be a brief procedure/ceremony whereby Catholic youngsters could become Muslim for short periods of time each day. This might be a handy solution for Catholic theologians providing pastoral guidance to youthful offenders.)

Given how difficult it can be to sort out the meaning of statements that are seemingly at odds with one another, it is not surprising perhaps that a poll by the Pew Research Center's Forum on Religion and Public Life (December 9, 2009) finds that large numbers of Americans engage in multiple religious practices, mixing elements of diverse traditions. This is reinforced by a 2007 Pew report that shows 68 percent in the US say *There's more than one way to interpret the teachings of my religion.*

While practices are much more restrictive in some parts of the world, in the US at least those who are religious generally can choose not only which faith to belong to, but also which house of worship they will attend. (Of course, in a community with only one church you don't have much choice.) Within the constraints of their faiths, they also can determine for themselves which church teachings to embrace and which to ignore. Catholics who choose to ignore certain teachings are so numerous, in fact, that they are called *cafeteria Catholics* or *protest Catholics.* Likewise, they can choose which parts of the Bible to ignore and which miracles to believe. (Does anybody believe the story in Matthew 27:52 of zombies rising from tombs, by the way?)

Another Pew study (May, 2010) reports that about half of American adults have changed their religious affiliation at least once in their lives. Those who have left Catholicism outnumber those who have joined the Catholic Church by nearly a four-to-one margin. Pew reports that 48 percent of unaffiliated former Catholics left the Church because they were unhappy with its teachings on birth control. And that 67 percent of Catholics say their religion should either *adjust to new circumstances* or *adopt modern beliefs and practices.*

This is not just a US phenomenon, either. The BBC News described a 2011 study in Amsterdam that found that one-in-six clergy in the mainstream Protestant Church in the Netherlands and six other smaller denominations was either agnostic or atheist. (An **agnostic** is, broadly speaking, someone who is not committed to believing in either the existence or the nonexistence of God or a god. An **atheist**, on the other hand, is someone who does not believe in the existence of a god or any gods.) *To compete in this market of ideas, some Christian groups seem ready virtually to reinvent Christianity. . . . They believe that only through 'adaptation' can their religion survive.* (BBC News, August 5, 2011)

People abandon their faith traditions and choose others for many reasons, including disillusionment and missionary proselytizing. Another common reason is love. When the relationship of people of two different faiths seems headed toward marriage, it is often necessary or useful for one partner to agree to accept the other's religion. Decades ago the response of families and faith leaders to this was sometimes quite virulent, but in many parts of the world attitudes toward religiously mixed marriages have softened greatly outside of faiths that are extremely doctrinaire. In other places interfaith marriages evoke consequences that can be quite draconian, even deadly.

The right to choose your own religion based on your own beliefs is supported by the United Nation's Universal Declaration of Human Rights. (UN General Assembly in Paris on 10 December 1948)

Article 18. *Everyone has the right to freedom of thought, conscience and religion; this right includes freedom to change his religion or belief, and freedom, either alone or in community with others and in public or private, to manifest his religion or belief in teaching, practice, worship and observance.*

The freedom to express those opinions is also protected.

Article 19. *Everyone has the right to freedom of opinion and expression; this right includes freedom to hold opinions without interference and to seek, receive and impart information and ideas through any media and regardless of frontiers.*

Political election cycles and cultural, social, or religious hot-button causes seem to go hand in hand. Especially is this so in the United States, where politicians look to wedge issues to bring in donations and draw their kinds of voters to the polls, and religious leaders look to the election process as a way to increase their influence in city, state, and national government.

The issues change from cycle to cycle and sometimes virtually vanish from the scene in between, but politicians know that raising funds is easier when somebody has strong feelings about an issue, whether that is abortion/contraception, balanced budgets, creationism, economic justice, gays in the military, gun control, marriage equality, prayer in schools, terrorism/torture, voter fraud/voting rights, or welfare spending. When the campaign ends and the popular vote is taken, most move on to the next issue. For morality is not the absolute that religions routinely claim it is. The moral do's and don'ts of a religion usually are (1) not unique to that religion, and (2) still valid without the religion. As a society, we the people collectively decide what is moral.

Bill Donohue recently (August 31, 2015) reported in a Catholic League Survey of Catholics that 26 percent of the respondents selected, *I differ with the Catholic Church's position on some issues and the Catholic Church should modernize its beliefs by changing its position to reflect current public opinion.*

The issue of gays in the US military is an example of how opposition has been transformed into acceptance. First, scientific and cultural understanding of homosexuality changed. Then, in 2010, polls showed that 60 percent of Americans favored repealing the policy known as Don't Ask, Don't Tell, which didn't allow those serving in the military to be open about their sexuality. Next, after extensive debate, Congress repealed the law, the president signed the bill, and the military implemented it. The change has posed some problems for chaplains whose faiths consider homosexuality sinful, but a predicted exodus of chaplains protesting the change has not occurred. Nor has it affected military recruiting or unit discipline.

While top-down religious dogma can be overturned by the secular we-the-people democratic process, religious leaders have an advantage when communicating with the media and the public. If you or I told someone that we were getting messages from God, we likely would be watched closely and with suspicion. (If you speak to God, that's religious. If God speaks to you, that's schizophrenia.) If Pope John Paul II declares that the Virgin of Fatima deflected an assassin's bullet and thus saved his life, nobody asks for evidence. The statement enhances Fatima's standing among his followers, not to mention his own.

It is difficult for religions to admit that their interpretation of God's will has not only changed but also is continuing to change. Already Portugal, a long-time Catholic stronghold (and the home of Fatima), has made abortions and same-sex marriages legal. Ireland, where more than 84 percent of the population is Catholic, voted in 2015 to allow same-sex marriages. It will be interesting to see how long it takes for the Catholic Church to accept masturbation as not sinful and to allow gay marriages. The venerable British Catholic weekly *The Tablet* (March 2, 2013) argues, *There is no more mileage in this issue [gay marriage] for the Catholic Church, and the sensible course would be to put it on the back burner with the heat turned low—to make peace with the gay world and move on.*

Morality Outside of Religion

In all he says and does, man is obliged to follow faithfully what he knows to be just and right. (Catechism of the Catholic Church, CCC, 1749)

Morals are principles or habits with respect to right or wrong conduct. Where do morals originate? Is there moral grounding without God and religions?

Do onto others as you would have them do onto you. While many attribute the *Golden Rule* to the Bible (Luke 6:31), the ideal of reciprocity dates back hundreds of years before the New Testament. It is found in Confucianism, Hinduism, Buddhism, Taoism, Judaism, Zoroastrianism, and the rest of the world's major religions. It also can be found in in almost every ethical tradition. Philosophers such as

Pittacus and Thales promoted forms of it in ancient Greece. It does not require a god.

That has not prevented religions from attempting to co-opt morality in the same way that they took over Christmas, Easter, and weddings—secular events that became religious ones. Artificial *goodnesses* promoted by religions—devotions, religious celebrations, ceremonies, fasting, and pilgrimages—compete with, detract from, and substitute for actual helpful virtues.

Think for yourself. This common admonition, usually offered after your friends convince you to do something really stupid, also does not depend on God. Instead it emphasizes who ultimately is responsible for things when they happen.

Religious believers perceive that atheists are a dangerous influence. When the Pew Research Center asked Americans to rate a variety of groups on a *feeling thermometer* ranging from 0 to 100—where 0 reflects the coldest, most negative possible rating and 100 the warmest, most positive rating—Jews, Catholics, and evangelical Christians received an average rating of more than 60, atheists received an average rating of 41, and Muslims an average rating of 40.

Why are believers leery of those who do not believe in the existence of God? Because they think that without having to fear God people can—and will—do anything. Thus atheists must be evil or at least evil influences.

Then again, Catholic believers need only show a little repentance via the Sacrament of Penance or contrition at the last moment of life in order to be forgiven for something. They seemingly have a free *Avoid Hell* card. Other Christians are promised that their sins have been forgiven through the sacrifice of Jesus. This means that believers and non-believers alike can do just about everything with no bad supernatural consequences. Doesn't that give them more potential for doing harm?

In fact, numerous studies have shown that religious believers behave no better than non-believers. A 2014 project at the University of Illinois at Chicago, for instance, found that religious and nonreligious people commit similar numbers of moral acts. Religious America is significantly more crime-ridden than the more secular and peaceful Scandinavian countries, and crime rates in the religiously dominated South are notably higher than those in the more secular northeastern portion of the United States. In short, there is no evidence here that religion contributes anything to morality, though the University of Illinois study did note that religious people reported experiencing more intense self-conscious emotions—such as guilt, embarrassment, and disgust—after committing an immoral act than did nonreligious people.

Do humans care about what happens to others? Of course. Animals do, too. Altruism, empathy, and reciprocity are part of our human and animal natures. I once watched a rabbit that I was trying to catch as it circled outside a live-trap in frustration, apparently trying to help a panicked blackbird that had been caught in the trap instead. Animals also share, care, and exhibit kindness toward others. They generally accept and tolerate other animals and humans unless they feel threatened, only killing if they, their offspring, their food, or their territory appear to be at risk. Only the human animal kills for sport.

Obviously, nurture—our family, friends, and culture—contributes to the formation of our senses of morality as well as nature. Paul Bloom at Yale University is uncovering evidence that babies are born with the beginnings of compassion and empathy, attributes which may lay the foundation for future lessons in how to behave. If that's the case, is it really necessary to invoke God in order to teach a child not to hit a sibling with a toy? Or not to steal? Or not to lie? Do we need the threat of fire and brimstone to be good? A quote attributed to Abraham Lincoln suggests otherwise: *When I do good, I feel good, when I do bad, I feel bad, and that is my religion.*

We are taught early on that being good is better than being bad. The admonition to be good is not unique to any religion. It seems clear that we can have *faith* in the future of humanity, *hope* that all our lives get better, and exhibit *charity* towards our fellow citizens of the world, all while being independent of any particular religion.

Then Peter and the other apostles answered and said,
We ought to obey God rather than men. (Acts 5:29)

One of the basic tenets of democracy is that people are perfectly capable of establishing and living under laws of their own devising, rules of reasonable moral behavior that they create, as needed, through their community and secular governments rather than having them imposed upon them from some authority, political, military, or religious. It turns out too that when *we the people* establish laws for the common good, we regularly decide that some things are moral in spite of what it says in the Bible.

Slavery, supported in the Bible as noted earlier, was at one time considered both moral and legal in the United States, even by the US Supreme Court. Later, and with much turmoil, our civil society decided that slavery was immoral and should be abolished.

Prohibition of liquor was endorsed by a number of religions and organizations, which then pressed Congress and voters to pass an amendment to the Constitution that made it illegal to produce, transport, and sell alcohol. When eventually it was discovered that the majority of people did not support this attitude, the amendment, or the complications that arose from it, it was repealed.

A half-century ago US laws promoted by religions made it illegal to open places of business on Sunday, sell birth control devices, perform an abortion, or live together if unmarried. Laws that were intended to compel behavior deemed by one religion or another as moral have been overturned by the judicial branch of our government.

In cases where behavior demanded by civil law conflicts with behavior demanded by one's religion, civil law usually prevails. This is reflected in court cases centered on religious drug use (Arizona), medical care being withheld from children based on religion (Pennsylvania, Wisconsin), vaccinations being required (Ohio), non-discrimination in wedding and commitment ceremonies for same-sex couples (New Mexico), tithing to the church vs. alimony (Delaware), overturning an ordinance against religious funeral protests (Missouri), and the army's grooming regulations (Texas). Warren Jeffs, former president of the Fundamentalist Church of Jesus Christ of Latter-Day Saints, discovered recently that rape of under-age girls is not a protected religious rite, and he is now a convicted felon. The state of Louisiana is dealing with the conflict between mandatory reporting of sexual abuse of minors and sacred communications during a Catholic confession.

Civil law prevailing over religious freedom is also visible on the international scene. In 2012 the European Court of Human Rights (ECHR), in the case of *Staatkundig Gereformeerde Partij v. the Netherlands,* agreed with the Dutch Supreme Court that the *SGP* Dutch political party could not discriminate against women. The ECHR said in part: . . . *taken together the SGP's position is unacceptable regardless of the deeply-held religious conviction on which it is based.*

Until a few centuries ago, governments and religions could penalize people for perceived bad behavior. Today religions may influence the making of laws, but the legal system in the US usually abides by those laws as if the religions did not exist. In fact, if religions suddenly vanished, the human-generated moral codes embedded in our civil and criminal laws and regulations would continue to prevail.

There's a big difference, of course, in the punishments meted out by political and religious authorities. Not only can religions subject offenders to earthly sanctions such as excommunication and shunning, but they can also claim to penalize them in the *next life.*

Today, for instance, virtually all Christian churches preach that God consigns to everlasting Hell those who do not believe the Bible story of how God created the universe, died on the cross, and rose from the dead. What about the billions of Buddhists, Hindus, Muslims, and others who have never heard these stories or have heard them and have chosen to believe otherwise? Do they go to Hell?

In either case, based on the descriptions given of Hell, eternal damnation seems out of proportion to the deed. Among other things, this means that God is running an inquisition worse than the Inquisition we humans abandoned centuries ago as being too cruel.

Morality does not require membership in anything.

Though it is wary of atheists and agnostics, the public seems to realize that being moral and being religious are not the same thing: *53 percent of the public says it is not necessary to believe in God in order to be moral and have good values. . . .* (Pew Research Center, June 24, 2014) Whenever you judge as not moral or a good human value something that a religion is advocating—flying planes into skyscrapers, withholding medical care (faith-healing), discriminating against someone because of their gender, race, or sexual orientation—you are demonstrating that your morality originates from within you rather than from within that religion.

So if morality originates within us, how can we objectively evaluate and sort the good from the bad? Recent scientific research on the origins of morality is dealing with the question: Is morality a rational behavior or is morality based on emotion and intuition, followed by rationalization?

When you found out the truth about Santa, did your behavior change from nice to naughty? Do you misbehave when the boss is absent? Do you cheat when nobody is looking? Or do you live as if you are always on some supernatural surveillance camera? Do you think you can observe a person's behavior and determine whether he or she is a religious believer or non-believer?

Just as there have been good theists (e.g., Sister Caroline and Father Casper, among the best of the teachers I had in grade school and college) and evil atheists (dictators Pol Pot and Stalin), there have also been evil theists (Pope Urban II, instigator of the First Crusade, and Pope Gregory IX, known for instituting the Papal Inquisition) and good atheists (composers Richard Strauss and Richard Rogers). Pol Pot and Stalin, who are together responsible for the deaths of more than 15 million people, clearly qualify as atheists and as evil. But their deeds are easily overshadowed by the genocide perpetrated by God, presumably a theist, in the story told in Genesis of Noah's ark.

Martin Luther, who criticized Catholicism and some of its practices, was excommunicated by the church, i.e., *[deprived] of all participation in the common blessings of ecclesiastical society*, including all sacraments except the Sacrament of Penance. Hitler, a baptized Catholic who mentioned God over thirty times in *Mein Kampf* and was responsible for the killing of millions, was never censured by the church—though the president of the Catholic League, Bill Donohue, claims otherwise.

Donohue says (March 24, 2015) that Hitler was excommunicated *lata sententiae*. This Latin phrase, meaning *sentence (already) passed*, is used in the canon law of the Catholic Church to indicate a penalty that follows ipso facto or automatically, by force of the law itself, when a law is contravened. The laws in question apply to an apostate from the faith, a heretic, or a schismatic; persons who throw away the consecrated Eucharistic species or take and retain them for a sacrilegious purpose; a person who uses physical force against the pope; a priest who uses confession as a pretext to solicit the penitent to break the commandment against adultery; a bishop who ordains as a bishop someone without a papal mandate and the person who receives the ordination from him; a confessor who directly violates the sacramental seal of confession; a person who procures a completed abortion; and accomplices without whose assistance a violation of a law prescribing *latae sententiae* excommunication would not have been committed.

My first question is this: Under which canon code was Hitler excommunicated? None of the above seem to fit. My second question: If the Catholic Sacrament of Confirmation *marks [an] individual forever as a Christian*, can this individual still be excommunicated *lata sententiae?* All this looks to me like an attempt at after-the-fact face-saving for something that should have happened, but didn't.

Sports often get credit for contributing to good health and physical fitness and teaching participants how to work with one another. They also are often complimented for helping players learn fairness, loyalty, tolerance, perseverance, trust in others, sharing, acceptance, and playing by the rules. At the same time, we regularly tolerate behavior in organized sports that on the street would result in arrest or civil fines.

One of the other things sports can inculcate is the win-at-all-costs attitude displayed in Leo Durocher's famous pronouncement that *Nice guys finish last.* We should not be shocked by the behavior of national newsmakers who seem to think that anything not specifically against the law is ethical, even moral, behavior. Some even go a step farther—anything is okay as long as one doesn't get caught. Maximize profit on Wall Street and in poker. If there's no law against it, it's moral behavior. Business ethics have been moved from personal responsibility to the responsibility of the IRS, the FCC, the SEC, and other regulatory agencies—at least to the extent that they are not prevented by deregulators from doing their work.

Many of these attitudes are first introduced to youngsters in their organized sports, which all seem, sometimes even in their earliest stages, to require referees. Players learn early that nearly anything is fair as long as the referee doesn't call it. Can you imagine the outcry that would erupt if a player, called safe by the ump when sliding into second base, stood up and announced that he actually was tagged before reaching the base and thus should be out? The rules of today's game don't allow a player to disagree with the umpire, unfortunately, and he would likely face the condemnation of his coach and teammates for his honesty if he did.

In the days of the pick-up games of yore, players had to negotiate among themselves to keep the game fair. They served as their own umps, regulating their own behavior. That is how we still operate in criminal trials: We turn to a jury of the accused's peers rather than to God or a panel of bishops to decide if it was murder or self defense when one person kills another. While sometimes mistakes are made, secular society is capable of deciding the morality/legality of serious issues and determining just punishments.

Prayer

We must also face the fact that certain attitudes deriving from the mentality of 'this present world' can penetrate our lives if we are not vigilant. For example, some would have it that only that is true which can be verified by reason and science; yet prayer is a mystery that overflows both our conscious and unconscious lives. . . . (CCC 2727)

Praying is an integral part of most religions. Prayers generally fall into one of three types:

A plea to God to do something differently than He intended to before hearing from you; asking a *higher power* to temporarily change a law of nature just for you

A statement memorized and repeated as indoctrination, a reminder of what to believe

Thanking and adoring God for His help and glory

Let's look at each of these, one by one.

Prayer is a plea to God to do something differently than he intended to before hearing from you; asking God to temporarily change a law of nature, just for you.

The Catechism of the Catholic Church states, *Prayer is the raising of one's mind and heart to God or the requesting of good things from God.* (CCC 2559) We are also led to believe that the more we pray the more prayer works and that it works better if more people pray together in a church. (There's strength in numbers, after all, and gathering everyone together makes taking up collections much more convenient.) The church says too that God likes musical prayer better than plain prayer, since it tends to *move more deeply the hearts of all in attendance and bring them closer to God, and prepare for God a feast of melody.* (YOUCAT 183)

Does an omniscient God need a prayer to realize there is an issue? Does he need advice from us mere unworthy mortals? After all, isn't praying for a medical recovery like telling God, *Let me call this one; my idea of whether I get well again is better than yours?*

Is God, perhaps, a little absent-minded and in need of a reminder of his great powers or who it is who deserves his favors? What better way to accomplish that than sending up a little advertisement intended to persuade him to deviate from what he intended—a miracle. (Sunday mornings must feel like one unending commercial for him rather than a day of rest. If God wants to truly enjoy a day off each week, he might want to invest in a good ad blocker.)

But when ye pray . . . your Father knoweth what things ye have need of, before ye ask him. (Matthew 6:7–8)

Furthermore, why pray if God already knows what you are going to ask for? Is it better to pray alone or in a crowded church? In words you craft yourself or words printed in the weekly bulletin that everyone is saying aloud and in unison? The Bible offers some answers in Matthew, but they don't exactly match current practice in Catholic or many other Christian services.

And when thou prayest, thou shalt not be as the hypocrites are: for they love to pray standing in the synagogues and in the corners of the streets, that they may be seen of men. Verily I say unto you, They have their reward. But thou, when thou prayest, enter into thy closet, and when thou hast shut thy door, pray to thy Father which is in secret; and thy Father which seeth in secret shall reward thee openly. (Matthew 6:5–6)

Are all prayers answered or only some of them? What the Bible says on the practice is very promising.

Therefore I tell you, whatever you ask for in prayer, believe that you have received it, and it will be yours. (Mark 11:24)

You may ask me for anything in my name, and I will do it. (John 14:14, also John 16:24)

Ask, and it shall be given to you. Seek, and you shall find. Knock, and it shall be opened to you. (Matthew 7:7)

And all things, whatsoever ye shall ask in prayer, believing, ye shall receive. (Matthew 21:22)

Oh, wait! Maybe not. There seem to be some exceptions to these rules.

You ask and do not receive, because you ask wrongly, to spend it on your passions. (CCC 2737, James 4:3) This seems like making a promise with your fingers crossed behind your back.

But let him ask in faith, nothing doubting: for he that doubteth is like the surge of the sea driven by the wind and tossed. For let not that man think that he shall receive anything of the Lord. (James 1:6-7)

Two frequent temptations threaten prayer: lack of faith and acedia . . . (CCC 2755) Oh, so if there was no response to your prayer or an answer that was different than the one you wanted, it's because you didn't believe strongly enough or you were suffering from a state of listlessness or torpor, of not caring or not being concerned with one's position or condition in the world? Just as in fairy tales, one has to believe in the magic for it to happen?

And what happens when Jesus/God prays? To whom is he praying? Himself?

And when he had sent the multitudes away, he went up into a mountain apart to pray: and when the evening was come, he was there alone. (Matthew 14:23)

And he withdrew himself into the wilderness, and prayed. (Luke 5:16)

It is easy to imagine that some prayers appear to have been answered: I prayed. What I asked for happened. Therefore the prayer (and God) gets credit. What about all the times I prayed and it didn't happen?

My praying, asking God for something, argues that God pulls the strings. It reinforces the notion that God, instead of me or the laws of nature, is in control of and responsible for my destiny.

Prayer can also be thought of as inaction or even laziness disguised as action. Instead of studying diligently for a test, we might pray instead to pass the test. Instead of planning rationally for the future, we could simply pray for a secure retirement. Instead of providing desperate, expensive, inconvenient, or nearly impossible medical attention, we pray for God to save the patient. In this situation prayer is a sign of desperation, as no other rational option is apparent.

No scientific study has ever supported the notion that prayer works, that it helps: In 2006 researchers in the largest scientific study conducted of third-party prayer, the *Study of the Therapeutic Effects of Intercessory Prayer (STEP)*, studied 1,800 patients recovering from coronary artery bypass graft surgery. The research team included members from Harvard Medical School and the famous Mayo Clinic. The investigators found no evidence of divine intervention.

There is no scientific evidence that a law of nature ever has been temporarily changed by prayer: In all human history there has never been a documented miraculous recovery in a case that is unequivocally unambiguous. Yes, at times people have claimed that someone has made an unexplained medical recovery. Conveniently, these tend to favor ambiguous cancer patients over amputees, persons with Down Syndrome or a congenital defect like cleft palate, or someone with paralytic polio. (Is it a miracle that you never see a miracle on YouTube?) In any case, upon learning about such an event, a commission of theologians—followed by a commission of cardinals and finally the pope—might investigate and then pronounce the remission of a disease as a miracle and credit it to a particular supernatural individual. When two or more of these miracles are on the record, the church then announces a new saint to the Catholic world.

Unintended consequences of prayer are usually ignored. Mark Twain, in *The War Prayer,* said, *Ponder this—keep it in mind. If you would beseech a blessing upon yourself, beware! lest without intent you invoke a curse upon your neighbor at the same time. If you pray for the blessing of rain on your crop which needs it, by that act you are possibly praying for a curse on some neighbor's crop which may not need rain and can be injured by it.*

Evil appears to come naturally to God; he does not require solicitous prayer to do deadly tragedies. Nobody could have asked God to punish Adam and Eve, for instance. According to the Bible, there was nobody else around to make the request. And nobody asked God to put a curse on a fig tree as Christ is reported to have done in Matthew 21:18–19.

God doesn't seem to need our prayer to *impose judgment* either. Pat Robertson, a Southern Baptist minister, says God wiped out the people of Haiti in an earthquake because He didn't like their pact with the devil. Evangelist Robert Jeffress has preached that the September 11 attacks were *God's punishment for abortion.*

Christian preachers and politicians accept divine wrath as common and normal. In fact, God could not respond to nor support a request to do bad, a curse, as this would make God a co-conspirator to evil. He can, however, watch over a plane or bus crash and then supposedly, as a result of someone asking for it in prayer, miraculously save the life of a child. He might do good if asked.

The Lord's Prayer (Matthew 6:12) counsels us to forgive, but God does not like to follow suit. For instance, He had the opportunity before the crucifixion to forgive Adam, Eve, and subsequently all of humanity, but He chose instead to kill his son.

The apostles appear to have been similarly short-sighted. Even though God delegated to them the power to forgive—*It was when he gave the Holy Spirit to his apostles that the risen Christ conferred on them his own divine power to forgive sins: 'Receive the Holy Spirit. If you forgive the sins of any, they are forgiven; if you retain the sins of any, they are retained.'* (CCC 976, John 20:22–23)—they also didn't use it to forgive Adam and Eve. As a result everyone who dies without being baptized, even infants who draw but a single breath, is condemned to Hell because of *original sin*. (Wait! I thought the sacrifice of Jesus fixed all that.) Think of all the suffering they and the church could have alleviated if the disciples had exercised this power more broadly. After all, *There is no offense, however serious, that the Church cannot forgive.* (CCC 982)

Before crossing the railroad tracks, do you pray there is no train coming or do you look both ways? In the face of a hurricane, tornado, fire, disease, or any other calamity, if one person faithfully prays and another scientifically plans for survival, who will be more likely to survive the catastrophe?

And likewise on the national and global level, which is more likely to thrive? A secular country that plans for its future survival by educating its citizenry about science-based reality and governs rationally? Or a country—say the United States—that sometimes tries to pretend it is a Christian nation and plans for its future based on faith-teaching, national days of prayer, and faith-based governing?

(Note: China is obviously secular, but an underground Christian movement is developing. Russian President Vladimir Putin said, *We are a secular state, of course, and cannot allow state life and church life to merge*, but Putin himself is reported to be a devout member of the Russian Orthodox Church.)

Prayer also serves as **a statement memorized and repeated as indoctrination, a reminder of what to believe.**

Anyone who's ever had an advertising jingle stuck in his head can attest to the success they can have in changing attitudes and behavior, especially if repeated endlessly. The more you see a slogan and say it or sing it, the more deeply ingrained it becomes.

Prayers like the Rosary with its Apostles' Creed, the Lord's Prayer, Hail Mary, and Glory Be to the Father keep reminding the praying person to believe they are dependent on help from a higher power and what to believe about that power. What else could be the purpose of repeating the same prayer 53 times within a few minutes other than indoctrination, burning in those ideas? Is saying a prescribed set of words more often or more loudly, earnestly, fervently, melodically, or collectively supposed to be more effective with God? Didn't God get it the first time?

Prayers can serve as mob psychology, like a cheering section, providing a sense of togetherness.

I myself have often encouraged the frequent recitation of the Rosary. . . . The Rosary is also a path of proclamation and increasing knowledge, *in which the mystery of Christ is presented again and again at different levels of the Christian experience. Its form is that of a prayerful and contemplative presentation, capable of forming Christians according to the heart of Christ. . . . The Rosary retains all its power and continues to be a valuable pastoral resource for every good evangelizer.* (Pope John Paul II, ON THE MOST HOLY ROSARY, Apostolic Letter, February 2003)

. . . the Rosary is a school of faith! (Pope Francis, Vatican City, October 6, 2013)

Praying to our Father should develop in us the will to become like him and foster in us a humble and trusting heart. (CCC 2800)

Prayer internalizes and assimilates the liturgy during and after its celebration. (CCC 2655)

The memorization of basic prayer offers an essential support to the life of prayer, but it is important to help learners savor their meaning. (CCC 2688)

And to reinforce the importance of prayer:

Those who pray are certainly saved; those who do not pray are certainly damned. (CCC 2744)

It should be obvious why Pope Paul II wanted Catholics to rediscover the Rosary. It consists mainly of the following four mind-moulding prayers interspersed with mention of five mysteries from one of the following four sets—Joyful, Luminous, Sorrowful, and Glorious—surrounding the story of Jesus.

The Apostles' Creed got its name from the fact that it was believed to have been jointly created by the apostles under the inspiration of the Holy Spirit, with each of the twelve contributing one of the articles, but its wording is arguably more a product of the 4th century CE than the first. Other prayers offer similar statements—for instance, the Niceno-Constantinopolitan or Nicene Creed, the only authoritative ecumenical statement of the Christian faith accepted by the Roman Catholic, Eastern Orthodox, Oriental Orthodox, Anglican, and major Protestant denominations. It *draws its great authority from the fact that it stems from the first two ecumenical Councils (in 325 and 381).* (CCC 195) The Councils of Toledo, Lateran, Lyons, and Trent also produced their versions of the Catholic creed. *We shall find the chief truths which the Church teaches in the Apostles' Creed.* (The Baltimore Catechism of 1891)

[Caution: Read the following at your own risk. Don't blame this author if you end up being converted.]

Apostles' Creed
I believe in God, the Father Almighty, Creator of heaven and earth,
and in Jesus Christ, His only Son, our Lord,
who was conceived by the Holy Spirit, born of the Virgin Mary,
suffered under Pontius Pilate, was crucified, died and was buried;
He descended into hell; on the third day He rose again from the dead;
He ascended into heaven, and is seated at the right hand of God the Father Almighty;
from there He will come to judge the living and the dead.
I believe in the Holy Spirit, the holy Catholic Church,
the communion of saints, the forgiveness of sins,
the resurrection of the body, and life everlasting. Amen. (YOUCAT 28)

According to the New Testament, the Lord's Prayer, also called the Our Father and the Pater Noster, was taught by Jesus to his disciples. (Matthew 6:9–13) *The Lord's Prayer is truly the summary of the whole gospel. . . .* (CCC 2761; 2774), (YOUCAT 514)

The Lord's Prayer (repeated six times):
Our Father who art in heaven, hallowed be thy name.
Thy kingdom come, thy will be done on earth as it is in heaven.
Give us this day our daily bread, and forgive us our trespasses,
as we forgive those who trespass against us;
and lead us not into temptation, but deliver us from evil.
For thine is the kingdom and the power and the glory forever. Amen. (YOUCAT 511)

The Hail Mary, also commonly called the Ave Maria, asks for the intercession of the Virgin Mary. The prayer appears to date back to about 1050. (By the way, if Mary is the Mother of God, who created Mary?)

Hail Mary (repeated fifty-three times):
Hail, Mary, full of grace, the Lord is with you.
Blessed are you among women,
and blessed is the fruit of your womb, Jesus.
Holy Mary, Mother of God, pray for us sinners,
now and at the hour of our death. Amen. (YOUCAT 480)

Glory Be to the Father, also known as the Gloria Patri, is a doxology, a short hymn of praise to God.

Glory Be to the Father (repeated six times):
Glory be to the Father, and to the Son, and to the Holy Spirit, as it was in the beginning, is now and ever shall be, world without end. Amen. (YOUCAT 481)

Every prayer is followed with *Amen*, which is an expression of agreement with what has been said—an affirmation of faith.

Prayer can also be a means of **thanking and adoring God for His glory and help.**

Thank you, dear Lord, for this bountiful meal. It's standard practice at many club luncheon meetings and holiday dinners for someone to offer an invocation before the event begins. I always wonder: Did the designated prayer leader stand and thank God out loud for a fabulous dinner yesterday? Shouldn't the farmer, food distributer, wait staff, and cook get some credit for this meal? Does this person offer this prayer in this style when he's eating alone? Why is this prayer primarily deemed necessary when the audience is a captive one, composed, potentially, of both believers and non-believers? Couldn't everyone do an invocation quietly in his or her own way, according to his or her own beliefs? Why does everyone at the gathering feel the need to join in or quietly tolerate the practice?

Then there's the prayer itself. In the case of this particular one, there are only two possibilities regarding God's role in supplying the meal: Either He did or did not contribute to it. If He did not contribute, this prayer is a pointless distraction. If He did somehow contribute something, then He deserves credit. But . . . What is God's contribution to a lousy meal? Or to the deficient meals of the millions, young and old, who are starving in this world? Does God only get credit for abundant food and no blame for food shortages? If later it is discovered that this fine meal included E. coli, does God get to keep the credit?

Evangelist Billy Graham says that he has to *accept, by faith, that God is sovereign, and He is a God of love and mercy and compassion in the midst of suffering.* In response to which I ask, *How could an omnibenevolent god withhold food to the point of starvation or driving a desperate person to kill another for something to eat? Doesn't this sound a lot like flying reindeer or a cruel scheme of some sort?*

Illogic Nests

Remember back in your school days how enthusiastically you sang your school song and yelled your favorite cheers at athletic contests with neighboring schools? The main themes were *Go Team!* and *We Are The Best!* The problem is: The members of the opposing school thought they were the best too.

And it goes beyond school. As we move on in life, we want to believe we live in the best city in the best nation. That we have the best doctor and dentist. That our political party is the country's best hope. That our team deserves to win the championship. (If they stop winning, though, we stop going, and if we move to a new town, we might switch allegiances. There can be tremendous pressure to support the local team.) We also want to believe we belong to the best religion, the *true* one.

Over time, even though they're not perfect, we get comfortable with our doctor, dentist, professional ball team, political party, and especially our religion and we are reluctant to change. We settle into what I call *illogic nests* because they give the illusion, the feeling of logic, and they cradle us. They protect us. They help identify us. We understand them, and so we stick with them. It's like reluctantly spending more on an old jalopy because we've invested so much in it already. It's easier to stay in that nest than make a big dangerous leap to another nest.

People often are born into a particular religion nest, generally that of their parents. They get baptized and confirmed, maybe even educated in a church school. Leaving the nest is scary. Leaving the nest can also set off extreme efforts by others to persuade them to stay, efforts that can extend to threats of being disowned, in some religions such as Islam, even to death.

> *All the deeds of the apostate become null and void in this world and the next. He must be killed.* (An interpretation of: Surah II. 214 by Al Khazan and used in the Muhammadan University called Al Azhar)
>
> *We shall cast terror into the hearts of those who disbelieve.* (Qur'an 3:151)

Why does the United States tolerate Islam or any other religion that terrorizes or threatens to kill those that stop believing the tenets of a religion? Are the rights and privileges of a religion really that absolute and untouchable? If the author or reader here threatened to terrorize or kill someone, for any reason, we would not long remain free.

Who decides which religion is the correct one? What if your grandparents *jumped the nest* and landed in a different religion which their descendants are now following? If you die in this new religion and find yourself in Hell, who is to blame—you or your grandparents?

There is comfort and security in numbers, a unifying effect. There is satisfaction in knowing what others in the nest are probably thinking. (But note: Some Republicans support public education, abortion, and Social Security, while some Democrats support free enterprise, smaller government, and private schools. Some Catholics support birth control and evolution.)

If we lived in another city, we might very likely be rooting for a different team. If we had grown up in different economic or social conditions, we might reside in a different political nest. And if our parents had followed a different religion, we might think that their faith was the only one that made sense.

Let's next take a closer look at a significant religion, a popular faith tradition today.

Part III

Christianity

Whoever would become a Christian must first tear reason from the mind. (Martin Luther)

Christianity is about Jesus Christ, the son of God, and his sacrificial death and resurrection for us, for our salvation into everlasting life. John 3:16, the Bible's most commonly quoted verse, describes it this way: *For God so loved the world, that he gave his only begotten Son, that whosoever believeth in him should not perish, but have everlasting life.*

Or, as others have said: *God sacrificed Himself to Himself to save us from Himself.* (Google, Bing)

According to the God Premises, by the way, this shouldn't have been necessary. God is omnipotent, omniscient, omnipresent, and omnibenevolent. We aren't, but we can forgive someone without killing someone else. Can't God? Why was the giving (killing) of his son a necessary part of his forgiveness of humanity? Why isn't this verse simply: *For God so loved the world that whosoever believeth in him should not perish, but have everlasting life?*

Early History: In the (Very) Beginning

In the pre-Christian era of a few thousand years ago there were many gods with interesting, mysterious lives and unique responsibilities. The stories of these gods are now viewed as myths, but at the time, they were passed along as descriptions of reality. They included:

Gods born of a virgin: Adonis, Attis, Ra, Horus, Krishna, Mithra, Perseus, Romulus, Remus, Mars, Osiris, Hercules

Virgin mothers of gods: Neith, Anahita, Isis, Semele, Juno

Gods born on or near December 25 (at the time of the winter solstice): Adonis, Dionysus, Mithra, Osiris, Attis

Gods crucified: Agni, Dionysus, Ishtar, Krishna, Osiris, Vedas, Mithra

Gods resurrected: Adonis, Dionysus, Semele, Osiris, Attis, Mithra

Gods that performed miracles: Adonis, Mithra, Horus, Dionysus

One can easily see the Jesus story is not very original. It contains mysteries and myths from a variety of cults and religions, combined with elements specific to Christianity, all of which aim at understanding the unknowable via faith.

Now faith is the substance of things hoped for, the evidence of things not seen. (Hebrews 11:1)

Monotheism appeared about 600 BCE, but there were still many other gods.

Thus saith the Lord the King of Israel, and his redeemer the Lord of hosts; I am the first, and I am the last; and beside me there is no God. (Isaiah 44:6)

So if you're not first, God, you're nothing?

Now I know that the Lord is greater than all gods: for in the thing wherein they dealt proudly he was above them. (Exodus 18:11)

Among all the gods, the new Christian god was top god.

Mithra, a Persian god, was widely venerated by Romans centuries before the time of Jesus. Mithra was born around December 25, of a virgin mother Anahita, met three Wise Men, died on a cross, had twelve disciples, and ascended into heaven. Mithra was an early popular competitor of Christ. *By the third century CE Mithraism and its mysteries permeated the Roman Empire and extended from India to Scotland.* (D. M. Murdock, author of *Was Mithra Born of a Virgin Mother?* and *Who Was Jesus?*)

In the Roman Empire of the first century CE there were many miracle workers and crucifixions. Imagine a countryside swarming with famous, popular preachers like Jerry Falwell, Billy Graham, Fred Phelps, Andrew Young, Diane Knippers, Jimmy Swaggart, Pat Robertson, Creflo Dollar, T.D. Jakes, Ralph Winter, Richard Land, Kenneth Copeland, and others, each collecting their own followers and most claiming inspiration by God. These were not all mainline religion preachers. Many were non-denominational. Great speakers and storytellers, they found success with new ideas that eventually coalesced into Christianity.

In the early days Christianity was viewed as a cult and, just like today, cults were perceived and persecuted as a threat to the larger established religions. Christianity took root in many places, however, with bishops in Antioch, Rome, Corinth, Sardis, Alexandria, Jerusalem, Ethiopia, Damascus, Constantinople, Caesarea, etc. As it evolved, these disparate groups developed their dogmas and grew as an organized religion and these early independent bishops gained regional political power and influence.

It was a slow process. As noted, Mithraism was the dominant religion in the Roman Empire. The Romans also honored their emperor as a god. Some Roman coins in the *Jesus* era had Caesar Augustus (63 BC–14 AD) on one side with *divi filius* (Son of God) on the reverse side.

In 64 CE Rome was destroyed by a fire. Nero blamed the Christians, which allowed the state to seize Christian properties.

Roman Emperor Constantine I converted to Christianity and in 313 CE issued an edict of toleration, which made Christianity lawful. He called the Council of Nicaea in 325. He didn't want splits in Christianity to split the empire. Now all the bishops could get on the same page. In addition to creating the first uniform Christian doctrine, the Nicene Creed, the council's main accomplishment was settlement of the nature of the Son of God and his relationship to God the Father: the bishops declared him **true man and true God**, *begotten, not made*, brought into being *from the substance of the Father.*

Does it seem strange that it took 300 years to firm up the story that Jesus really was God? Well, consider what the bishops had to work with. Many contemporaneous accounts confirm the existence of Julius Caesar. No documents of the period confirm the existence of Jesus Christ, even though he performed miracles and rose from the dead. In short:

All early Roman sources about Christianity depend on Christian sources.

There is no record of anything having been written *by* Jesus Christ.

There is no record of anything having been written *about* Jesus Christ during his lifetime.

By the Edict of Milan (380 CE), Emperor Theodosius I made Christianity the official religion of the Roman Empire and granted religious freedom to all religions. This fostered Christianity in Rome. Soon after, in 391 CE, a Theodosian decree banned all pagan rites. Mithraism withered and went underground. After this, Christianity thrived with added credibility going to the Roman bishop. Gradually Roman bishops gained more power and influence and Rome became the power center of Christianity. By the 5th century the bishop of Rome came to be called pope. Today the pope is still called The Bishop of Rome.

The Greeks and Romans made many advances in the sciences and arts during their heydays, but that seemed to change as Christianity took over Rome in the 4th century. The next thousand years, during which Christianity dominated Western civilization, are known as the Dark Ages. During the following Medieval period, which spanned a few centuries, it became possible to begin to question, at great risk, Catholic control and dogma. The Age of Reason then followed, with science showing how to understand reality in a way religion never could.

By the way, China today might provide a window on early growth of Christianity in Rome. The rapid rise in the number of Christians in Europe from the 3rd to the 15th centuries may now be duplicated in China in only one century. The persecution that took place there under Mao is slowly giving way to acceptance. By one estimate, by 2030 the number of Chinese Christians could reach 250 million—the largest Christian population of any country in the world. (See The Economist, November 1, 2014)

The God Story

There are variations of the God story within the Christian community and elsewhere. Mine goes something like this:

After God created a universe with two humans, Adam and Eve, He decided that it would be better if they didn't know anything about good and evil. God told them not to eat an apple from the tree of knowledge of good and evil, even though—being omniscient—He knew full well they would eat the apple and what would happen if they did.

Being more than a little naïve, Adam and Eve ate the apple, since *evil* did not yet exist in their garden and apparently nobody told them not to listen to talking snakes. And for this disobedience, this *original sin*, God put a curse of sin, death, and eternal damnation not just on them, but on all of humanity forever after.

When His experiment in world-making continued to go badly, God decided he had to start over. Drowning every living thing that he earlier created on Earth—except for a few people and creatures on an ark—worked for a while, but eventually . . .

God decided He had to do something directly to help the sinful humans He had created. He then impregnated his mother with himself—by means of a mystical version of himself called the Holy Spirit. (The result is Jesus Christ, who is a human with a mother but who is his own father.) The son of God tried continually to prove to the world that he, Christ, really was their god. If they believed his story, they would be absolved from an abrupt termination and live with him happily ever after. After about 33 years, though, God had to deal again with the fact that the humans he created were still behaving like humans; they retained the sinful nature that he gave them and often needed forgiveness.

It turned out, unfortunately, that they couldn't join him and adore him after they died unless he sacrificed his innocent self to himself to atone for the innocent, but bad act of somebody long ago disobediently eating an apple. In a spectacular agonizingly torturous and prophesied bloody *Suicide by Cop* finale, Jesus gets killed on a cross by Roman soldiers while praying to himself. A few days after that famous sacrifice, as a grand proof that the sacrifice was over and he might have been kidding, but he really was a god, God raises himself (well, Jesus) from the dead—something which obviously every good God should be able to do—and shows off what He did.

The Bible story ends where Jesus joins himself and the Holy Spirit (also himself?) in Heaven and lives happily ever after with all the humans who have been saved by believing this story.

Remember, to be saved, to join him (them?) after you die, you must believe this story of his life, death, and resurrection. You must have faith that this is a true mystery, as Christian enterprises are forever reminding us:

. . . Believe on the Lord Jesus Christ, and thou shalt be saved . . . (Acts 16:31) *Believing in Jesus Christ and in the One who sent him for our salvation is necessary for obtaining that salvation.* (CCC 161)

He that believeth on the Son hath everlasting life: and he that believeth not the Son shall not see life; but the wrath of God abideth on him. (John 3:36)

That if thou shalt confess with thy mouth the Lord Jesus, and shalt believe in thine heart that God hath raised him from the dead, thou shalt be saved. (Romans 10:9)

He that believeth on him is not condemned: but he that believeth not is condemned already, because he hath not believed in the name of the only begotten Son of God. (John 3:18)

Each Christian community is called to be a place where people can be taught about the mysteries celebrated in faith. (Pope Benedict XVI)

There's that *faith* thing again. It brings to mind this exchange in *Alice in Wonderland*:

There's no use trying, Alice said. *One can't believe impossible things. I daresay you haven't had much practice*, said the Red Queen. *When I was your age, I always did it for half-an-hour a day. Why, sometimes I've believed as many as six impossible things before breakfast.*

Is what she is saying really all that different from the following statement in the *Youth Catechism of the Catholic Church*?

We see in the life of Jesus numerous realities that are powerfully present but that we can understand only as a mystery. Examples of such mysteries are the divine Sonship, the Incarnation, the Passion, and the Resurrection of Christ. (YOUCAT 78)

How does one *understand only as a mystery* something that is not, by definition, understandable? That must be why it's called a mystery.

The God story provides many opportunities for believing impossible things. So do many of the official pronouncements about it, which are often contradictory. Consider this statement in the *Youth Catechism of the Catholic Church*:

Are we responsible for the sins of other people? *No, we are not responsible for other people's sins.* (YOUCAT 319)

The fathers shall not be put to death for the children, says the Bible, *neither shall the children be put to death for the fathers: every man shall be put to death for his own sin.* (Deuteronomy 24:16) So, why were the descendants of Adam and Eve cursed for their transgressions? What did the plants, animals, and innocent children of the world do to deserve death in the flood? And why did Jesus volunteer to be crucified for something he didn't do? (Think about it—Christianity has as its core belief the torturous death of an innocent man.)

Besides all that—how can the suffering of another excuse something you have done? And if it could, do you think it right and proportionate for someone else to die to atone for Adam and Eve disobeying God?

Here's a thought: What if God had shown a little more understanding in the first place and not cursed Adam and Eve and all their descendants? Or if, after He thought about it for a while, He removed the curse? Why did He cast the spell initially if He knew he was going to have to kill his son to undo it? Jesus didn't do anything wrong.

If some guy dissed you and ate one of your apples, could you/would you kill your only son to save that guy from an eternal damnation that you initiated? Christianity essentially says *yes*. After all, that's what God did, didn't He?

Set Up to Fail

God encourages the survivors of the flood to be very fecund. *And God blessed Noah and his sons, and said unto them, Be fruitful, and multiply, and replenish the earth.* (Genesis 9:1) Whatever they do, though, none of these descendants can save themselves from the everlasting hell he consigned them to when he cursed Adam and Eve.

How are we saved? *No man can save himself. Christians believe that they are saved by God, who for this purpose sent his Son Jesus Christ into the world . . .* (YOUCAT 337)

God couldn't avoid this with a *curse reverse* from Heaven? He can't overrule himself? Well, at least he gives us a way out—believing in him. Even so, even with the assistance of his son Jesus, God is only able to help a small portion of us avoid perdition (hell):

Enter through the narrow gate. For wide is the gate, and broad is the way, which leads to perdition, and many there are who enter through it. How narrow is the gate, and how straight is the way, which leads to life, and few there are who find it! (Matthew 7:13–14)

So, it appears that after God's vast reset, the great flood, humans are still set up for failure. All continue to be *fallen* by default, but a select minority, the *believers*, can be saved. Think about it: God created a situation where less than half are saved. A minority are successful. By God's own reckoning. God had to know beforehand how this was going to play out. Should Christians be happy with being bushwhacked by a god that creates people that he is expected to save but who, he knows, will primarily end up in perdition?

The dividing line between heaven and hell is infinitesimally narrow. And you'll never know how you did until you die. It's like a championship basketball game with virtual foul boundaries, virtual out-of-bounds lines, baskets, clock, and scoreboard. The ref has magic glasses and, of course, can see everything. You play your heart out until all of a sudden the game is over and it's time to find out if you won or lost. How long would you play this game under these rules? And what if it turned out you lost and as an attempt at consolation, the organizers said: *Well, you did have 'free will.'*

When God says, *Love me, believe in me, and accept me as your savior, or suffer in hell and, by the way, deciding what to do is totally up to you—I've given you free will*, it's like a crook placing a gun to your head and saying, *Give me your money or die*, and claiming it's your choice. (If you choose not to give up your money and get shot, does that make your death a suicide rather than murder? Is the crook innocent?)

Abuse and Christianity's God

Domestic violence is a pattern of abuse that a significant other uses to control the behavior of another person. There are many common, recognized symptoms of abuse. Do any of them match the behavior of Christianity's God?

The abused becomes emotionally and economically dependent. A Christian only gets *grace* from God, which allows one to strengthen one's faith in God.

Between God and us there is an immeasurable inequality, for we have received everything from him, our Creator. (CCC 2007)

Every good gift and every perfect gift is from above, and cometh down from the Father of lights. . . . (James 1:17)

Second, material poverty can be an occasion to draw greater attention to the spiritual needs of the human person, thus leading to a deeper reliance on the Lord, from whom all good things come. (Pope Francis, Vatican City, September 24, 2014)

Man is a beggar before God. (St. Augustine, CCC 2559)

The abused irrationally thinks the abuser is everywhere and knows everything. Virtually all Christians accept that their God is omnipresent and omniscient.

The abuser is jealous. Yes, very.

I am the Lord, your God. . . . You shall have no other gods before me, for I the Lord thy God am a jealous God, visiting the iniquity of the fathers upon the children to the third and the fourth generation . . . (Exodus 20:2–5)

He that loveth father or mother more than me is not worthy of me: and he that loveth son or daughter more than me is not worthy of me. (Matthew 10:37)

The abuser does not trust the abused. God is everywhere and is always watching and judging you. You're never really alone.

The abused gets isolated from family and friends. Your family and friends can't help you when it comes to God. You are on your own. Apostasy also often brings isolation.

And every one that hath forsaken houses, or brethren, or sisters, or father, or mother, or wife, or children, or lands, for my name's sake, shall receive an hundredfold, and shall inherit everlasting life. (Matthew 19:29)

The abuser is overly possessive. Only God can tell you what is good and bad. You may beg and pray for your God to help you. But only God decides. God created you. God tells you how to think and behave. God determines when you die. God wants you to live with and adore him forever. God owns you.

The abuser threatens to harm the disobedient abused. Christians must follow God's rules, as interpreted by the preachers and churches, or go to hell. He made Adam and Eve into an example for all humans.

The abuser restricts resources from the abused. You only get enough of whatever Sunday services provide to get you to the next Sunday. If you had plenty of grace and faith, you wouldn't have to struggle to be saved. It could have been automatic, generous, sufficient, and guaranteed.

No man can gain heaven merely by his own efforts. (YOUCAT 341)

The abused is trapped and thinks there is no way to escape. Clearly the believer is trapped by baptism, confirmation, indoctrination, and the threat of excommunication for apostasy.

The abused are told that there isn't anyone else to blame for their abuse. It's their own fault. Christians are led to think they and all humanity beginning with Adam and Eve are the sinners who bring down God's wrath and thus need to be saved. It's your own fault if you go to hell, right? In fact, wasn't hell just made for you? You will suffer there more than the devil himself, who is not confined to hell. Isn't he all over our world trying to tempt you, suck you in?

For all have sinned, and come short of the glory of God. (Romans 3:23)

The abused is counseled not to provoke the abuser.

This [First] commandment forbids us . . . to provoke God by word or deed. (YOUCAT 355)

Good Christians, like slaves and soldiers, ask no questions. (Jerry Falwell)

The abused fears the abuser. Believers are told that there is no way to escape God's wrath and eventual punishment except by following God's instructions very carefully.

God's name is holy when said with veneration and fear of offending him. (CCC 2149)

Let us hear the conclusion of the whole matter: Fear God, and keep his commandments: for this is the whole duty of man. (Ecclesiastes 12:13)

The abused is made to feel inadequate. Believers are taught that everything good they achieve is due to God, not their own efforts.

For by grace are ye saved through faith; and that not of yourselves: it is the gift of God: Not of works, lest any man should boast. (Ephesians 2:8–9)

And said, Verily I say unto you, Except ye be converted, and become as little children, ye shall not enter into the kingdom of heaven. (Matthew 18:3)

We go to Mass because we are sinners and because we wish to receive God's forgiveness, to participate in Christ's redemption, his forgiveness. (Pope Francis, February 12, 2014)

[Jesus] calls us to follow Him on His path of humiliation. (Pope Francis, April 16, 2014)

To adore God is to acknowledge, in respect and absolute submission, the 'nothingness of the creature' who would not exist but for God. (CCC 2097)

Moreover, God wishes us to realize, and always keep in mind, that, unassisted by His heavenly grace, we can of ourselves do nothing, and should therefore apply ourselves to prayer with all the powers of our souls. (CCT)

By kneeling, a person makes himself small in the presence of God's greatness. He recognizes his dependence on God's grace. (YOUCAT 486)

As it is written, There is none righteous, no, not one. . . . (Romans 3:10)

For all have sinned, and come short of the glory of God. . . . (Romans 3:23)

Outward sacrifice, to be genuine, must be the expression of spiritual sacrifice: 'The sacrifice acceptable to God is a broken spirit.' . . . (CCC 2100)

Christianity teaches that the human race is depraved, fallen, and sinful. (D. James Kennedy, 1980)

But we are all as an unclean thing, and all our righteousnesses are as filthy rags; and we all do fade as a leaf; and our iniquities, like the wind, have taken us away. (Isaiah 64:6)

The abuser promises the abused is loved. *God loves you.* Ever hear or read that? He loves you so much he killed/sacrificed his only son for you. God still loves children so much that he allows only six million to die of starvation and neglect each year.

God is love; he is therefore righteous and faithful; he cannot contradict himself. . . . (CCC 2577)

God's love is the source, the support, and the goal of all holiness . . . (Pope Francis, VIS November 24, 2014)

If the abused dies, the abuser blames someone else.

God did not make death, and he does not delight in the death of the living . . . It was through the devil's envy that death entered the world. (Wisdom 1:13; 2:24)

Amazingly, another characteristic of domestic abuse is that the abused often wants to go back for more. The abuse is not even acknowledged. The abused cannot escape the abuse while denying its existence. Could it be that the big breakthrough for Christianity in the 1st-to-4th centuries CE was its institutionalization of these methods of abuse and linking them to threats on life or consequences after death?

Qualities of the Christian God: A Match Made in Heaven?

If I show you pictures of six different people, including one of Jesus, could you identify the picture of Jesus? In my experience, most people would say they could identify him. It's a trick question, of course. There is no physical description of Jesus anywhere in the Bible. Nobody knows what he may have looked like. Not Michelangelo of *The Last Judgment* fame. Not da Vinci, who painted him in *The Last Supper.* Not even Warner Sallman, who sold over 500 million copies of his *The Head of Christ* painting. Being able to recognize Jesus is like saying you could identify Santa Claus in a line-up. (You mean the real Santa Claus?)

We may not know what Jesus looked like, but the Bible tells us a great deal about what he and God are like as beings. Frankly, some of their attributes are not all that attractive. If you saw them in the profile of someone at an online dating service, they might give you pause. Are these the qualities of a God you think you could love and adore now and throughout eternity in heaven?

Anger issues

And the anger of the Lord was kindled against Moses. . . . (Exodus 4:14)

And when the people complained, it displeased the Lord: and the Lord heard it; and his anger was kindled. . . . (Numbers 11:1)

God judgeth the righteous, and God is angry with the wicked every day. (Psalms 7:11)

Thus shall mine anger be accomplished, and I will cause my fury to rest upon them, and I will be comforted: and they shall know that I the Lord have spoken it in my zeal, when I have accomplished my fury in them. (Ezekiel 5:13)

And there shall cleave nought of the cursed thing to thine hand: that the Lord may turn from the fierceness of his anger, and shew thee mercy, and have compassion upon thee, and multiply thee, as he hath sworn unto thy fathers. . . . (Deuteronomy 13:17)

Scripture reveals the presence of anger and envy in man, consequences of original sin, from the beginning of human history. (CCC 2259) It looks like man really *was* created in the image of God.

What does the fifth commandment forbid? The fifth commandment forbids murder and suicide, and also fighting, anger, hatred, revenge, drunkenness, reckless driving, and bad example. (Baltimore Catechism 253)

God does not appear to follow his own fifth commandment. And . . . isn't anger one of the famous seven *deadly* sins? Pope Francis appears to be unaware of this characteristic of his God: *One may anger at times; it is healthy to be angry at times. But the state of rage is not God's, and leads only to sadness and disunity.* (Pope Francis, Vatican City, July 28, 2014)

Believes in blood sacrifice

For God so loved the world, that he gave his only begotten Son, that whosoever believeth in him should not perish, but have everlasting life. (John 3:16)

So, it's morally okay to kill your son to prove how much you love someone else? Really? Christianity's core is built on this. Someone also needs to explain to me how a loving God could threaten and ultimately impose on humanity an infinite punishment for a finite infraction— disobediently eating an apple.

Biased

Even though he created them all, God appears to be prejudiced about women, homosexuals, and non-believers. (Not to mention pigs and shrimp.) He seems to favor Catholics, males, and the clergy.

Caring

Cast all your anxiety on him because he cares for you. (1 Peter 5:7)

Cruel

For I will show him how great things he must suffer for my name's sake. (Acts 9:16)

Blessed are ye, when men shall revile you, and persecute you, and shall say all manner of evil against you falsely, for my sake. (Matthew 5:11)

Allow your concubine to be gang raped all night, then cut her into a dozen pieces. (Judges 19:25-29)

He that spareth his rod hateth his son: but he that loveth him chasteneth him betimes. (Proverbs 13:24)

And if thy hand offend thee, cut it off: it is better for thee to enter into life maimed, than having two hands to go into hell, into the fire that never shall be quenched. . . . (Mark 9:43)

I form the light, and create darkness: I make peace, and create evil: I the Lord do all these things. (Isaiah 45:7)

One cannot name any human as evil as a God that would create an infinite punishment. A hell.

He thinks it's okay to whip your slaves.

And that servant, which knew his lord's will, and prepared not himself, neither did according to his will, shall be beaten with many stripes. But he that knew not, and did commit things worthy of stripes, shall be beaten with few stripes. For unto whomsoever much is given, of him shall be much required: and to whom men have committed much, of him they will ask the more. (Luke 12:47-48)

He thinks you shouldn't be nice to non-Christians.

If there come any unto you, and bring not this doctrine, receive him not into your house, neither bid him God speed: For he that biddeth him God speed is partaker of his evil deeds. (2 John 1:10-11)

Demanding (an extortionist)

According to the Catholic Catechism's interpretation of the first commandment as written by God, it commands one to love God or face eternal damnation for violating a commandment. Adam, Eve, and everyone after must discover the importance of obedience to God. Or else.

The first commandment summons man to believe in God, to hope in him, and to love him above all else. (CCC 2134)

This is like demanding, *You will love me, or go to hell.*

And why call ye me, Lord, Lord, and do not the things which I say? (Luke 6:46)

Divisive

I am come to send fire on the earth. . . . Suppose ye that I am come to give peace on earth? I tell you, Nay; but rather division . . . (Luke 12:49-51)

Easily provoked

This [first] commandment forbids us: . . . to provoke God by word or deed. (YOUCAT 355)

Even fig trees need to be careful:

Now in the morning as he returned into the city, he hungered. And when he saw a fig tree in the way, he came to it, and found nothing thereon, but leaves only, and said unto it, Let no fruit grow on thee henceforward for ever. And presently the fig tree withered away. (Matthew 21:18-19)

Fearsome

The fear of the LORD is the beginning of knowledge. (Proverbs 1:7)

Thou shalt rise up before the hoary head, and honour the face of the old man, and fear thy God: I am the Lord. (Leviticus 19:32)

Servants, be obedient to them that are your masters according to the flesh, with fear and trembling, in singleness of your heart, as unto Christ. . . . (Ephesians 6:5)

Genocidal

And all the firstborn in the land of Egypt shall die, from the first born of Pharaoh that sitteth upon his throne, even unto the firstborn of the maid-servant that is behind the mill; and all the firstborn of beasts. (Exodus 1:5)

God brags that He destroyed many nations. (FFRF, Freethought Today, March 2012)

Hateful

Of all the possible ways of creating a humankind, God chose a version that sends a majority to an everlasting suffering in hell. And he knew beforehand that would happen. Know God by what he does, not by what he says. God could have stopped with punishing Adam and Eve, but no, he had to curse and punish all of humanity.

As it is written, Jacob have I loved, but Esau have I hated. (Romans 9:13)

Not without good reason is hatred called the sin of the devil. (CCT)

But this thou hast, that thou hatest the deeds of the Nicolaitanes, which I also hate. (Revelation 2:6)

Helpless

God did not make death, and he does not delight in the death of the living. . . . It was through the devil's envy that death entered the world. (*Wisdom* 1:13; 2:24) (CCC 413)

So, God is helpless?

And God can't seem to stop Satan from doing what Satan does. Satan seems to a Christian like a competitor of God for your allegiance. And God even admits He is losing.

Enter through the narrow gate. For wide is the gate, and broad is the way, which leads to perdition, and many there are who enter through it. How narrow is the gate, and how straight is the way, which leads to life, and few there are who find it! (Matthew 7:13-14)

Holds grudges

God accepts generational sin, and imposes judgment upon the next generations.

Keeping mercy for thousands, forgiving iniquity and transgression and sin, and that will by no means clear the guilty; visiting the iniquity of the fathers upon the children, and upon the children's children, unto the third and to the fourth generation. (Exodus 34:7)

Thou shalt not bow down thyself unto them, nor serve them: for I the

Lord thy God am a jealous God, visiting the iniquity of the fathers upon the children unto the third and fourth generation of them that hate me. . . . (Deuteronomy 5:9)

Christianity claims all humans today suffer due to Adam and Eve's original sin. But humans are commonly expected to forgive and forget. Another instance of *Do as I say, not as I do?*

Inconsiderate

Christians believe God chooses when an individual dies without regard to that person's wishes. In fact, it is considered un-Christian for one to choose his or her own time of death, even if a long painful death is imminent. This attitude is evolving, with Oregon and California leading the way.

For I say unto you, That unto every one which hath shall be given; and from him that hath not, even that he hath shall be taken away from him. (Luke 19:26) Take from those that have not?

And another of his disciples said unto him, Lord, suffer me first to go and bury my father. But Jesus said unto him, Follow me; and let the dead bury their dead. (Matthew 8:21-22)

And another also said, Lord, I will follow thee; but let me first go bid them farewell, which are at home at my house. And Jesus said unto him, No man, having put his hand to the plough, and looking back, is fit for the kingdom of God. (Luke 9:61-62)

Does this sound like a cultish Jim Jones or maybe a David Koresh? And what happened to that commandment about honoring your father and your mother?

Lay not up for yourselves treasures upon earth, where moth and rust doth corrupt, and where thieves break through and steal . . . (Matthew 6:19)

Take therefore no thought for the morrow: for the morrow shall take thought for the things of itself. Sufficient unto the day is the evil thereof. (Matthew 6:34)

Bottom line: Don't save and plan for tomorrow, for tomorrow will take care of itself.

Jealous

Thou shalt not bow down thyself to them, nor serve them: for I the Lord thy God am a jealous God. . . . (Exodus 20:5)

He that loveth father or mother more than me is not worthy of me: and he that loveth son or daughter more than me is not worthy of me. (Matthew 10:37)

For thou shalt worship no other god: for the Lord, whose name is Jealous, is a jealous God. (Exodus 34:14)

Just

He is the Rock, his work is perfect: for all his ways are judgment: a God of truth and without iniquity, just and right is he. (Deuteronomy 32:4)

Loving

He that loveth not, knoweth not God; for God is love. (1 John 4:8)

Christ, the beloved Son, came to reveal the Father's love for humanity. Whoever hears the Gospel and responds to the call to conversion becomes Jesus' 'brother, sister and mother,' and thus an adopted son of his Father. (Pope Francis, Vatican City, December 10, 2014)

Recall, please, that God's innocent son was sacrificed on the cross to show the father's love for humanity. According to the pope, you could be an adopted child of God the Father. Let's hope he would be nicer to you, an adopted relation, than he was to his own son, Jesus.

Misogynistic

Let your women keep silence in the churches: for it is not permitted unto them to speak; but they are commanded to be under obedience as also saith the law. And if they will learn any thing, let them ask their husbands at home: for it is a shame for women to speak in the church. (1 Corinthians

14:34-35, with related comments at Leviticus 12:1-8, Leviticus 15:19-31, Leviticus 27:1-7, Exodus 18:21, Exodus 20:17, Exodus 21:7, 1 Timothy 2:11-15, 1 Corinthians 11:2-10, Deuteronomy 1:13, Deuteronomy 22:20-24, Deuteronomy 22:28-29, Matthew 25:1, John 20:17-27, Genesis 2:18, Genesis 3,16, Ephesians 5:22-24, Numbers 11:24, Numbers 30:1-16, Numbers 31:15-18, Revelation 14:3-4, Matthew 5:31-32, 1 Kings 11:3, Colossians 3:18, Luke 2:22-23, and Luke 11:27-28)

Note: Every book in the Bible apparently was written by a man. However, Christian misogyny is not, by any means, limited to the Bible:

Woman was merely man's helpmate, a function which pertains to her alone. She is not the image of God but as far as man is concerned, he is by himself the image of God. (Saint Augustine)

Any woman who acts in such a way that she cannot give birth to as many children as she is capable of, makes herself guilty of that many murders. (Saint Augustine)

Women should remain at home, sit still, keep house, and bear and bring up children...If a woman grows weary and, at last, dies from childbearing, it matters not. Let her die from bearing—she is there to do it. (Martin Luther)

Long story short: Women—don't talk in church. If you have a question, ask your husband at home. Thankfully, this attitude largely has evolved to where it died out except in fundamentalist faiths.

Needy

Regarding your obligation to always pray:

You cannot keep God content with a few words in the morning or evening. (YOUCAT 510)

It's our obligation to keep God content?

We must, then pray in spirit and in truth; for the heavenly Father seeks those who adore Him in spirit and in truth. (CCT)

Apparently God can't get enough adoration.

For we have not been commanded to work, to keep watch and to fast constantly, but it has been laid down that we are to pray without ceasing. (CCC 2742)

For this forms faith.

Omnipotent

For with God nothing shall be impossible. (Luke 1:37)

Well, *almost* nothing is impossible. *And the Lord was with Judah; and he drave out the inhabitants of the mountain; but could not drive out the inhabitants of the valley, because they had chariots of iron.* (Judges 1:19) Was the writer referring to the inhabitants of the valley having a military infrastructure too advanced for God to handle? Or was it because the chariots were made of iron, which historically is believed to repel, contain, or harm ghosts, fairies, witches, and/or other malevolent supernatural creatures?

Of all the divine attributes, only God's omnipotence is named in the [Nicene] Creed. (CCC 268)

Peacemaker

Now the God of peace be with you all. Amen. (Romans 15:33)

Plunderer

And when the Lord thy God hath delivered it into thine hands, thou shalt smite every male thereof with the edge of the sword: But the women, and the little ones, and the cattle, and all that is in the city, even all the spoil thereof, shalt thou take unto thyself; and thou shalt eat the spoil of thine enemies, which the Lord thy God hath given them. (Deuteronomy 20:13-14)

And the Spirit of the Lord came upon him, and he went down to Ashkelon, and slew thirty men of them, and took their spoil, and gave change of garments unto them which expounded the riddle. And his anger was kindled, and he went up to his father's house. (Judges 14:19)

Regretful

And it repented the Lord that he had made man on the earth, and it grieved him at his heart. (Genesis 6:6)

So why did he do it? Why didn't he start over? Too lazy? Not enough time?

Rigid

For I am the Lord, I change not. (Malachi 3:6)

Jesus Christ the same yesterday, and today, and for ever. (Hebrews 13:8)

Be ye therefore perfect, even as your Father which is in heaven is perfect. (Matthew 5:48)

Sadistic

Sadism is a behavioral disorder characterized by callous, vicious, manipulative, and degrading behavior expressed towards other people.

But those mine enemies, which would not that I should reign over them, bring hither, and slay them before me. (Luke 19:27)

Of course, God can't kill his enemies himself. Let others do it.

Unto the woman he said, I will greatly multiply thy sorrow and thy conception; in sorrow thou shalt bring forth children; and thy desire shall be to thy husband, and he shall rule over thee. (Genesis 3:16)

Is there no end to the punishment threatened by God? God finds it easier to kill than to forgive.

Senseless

God talks only to believers, the faithful. He *preaches to the choir.* After thousands of years he has won over a fourth of the world's population. Imagine how things would change if he started talking convincingly to non-believers, to atheists. He could have everyone on his side in no time flat.

Look again at the story of Adam and Eve. After they had eaten the apple, Adam and Eve hid behind a tree where somehow the omniscient God couldn't find them.

And they heard the voice of the Lord God walking in the garden in the cool of the day: and Adam and his wife hid themselves from the presence of the Lord God amongst the trees of the garden. And the Lord God called unto Adam, and said unto him, Where art thou? (Genesis 3:8-9)

This is obviously the universe's first game of hide-and-seek or possibly God's first attempt at humor. Not being able to find Adam and Eve is not very impressive for an all-knowing God. Nor is him having to ask Adam and Eve if they ate an apple.

And he said, *Who told you that you were naked? Have you eaten from the tree of which I commanded you that you should not eat?* (Genesis 3:11)

Robert Ingersoll wrote in *Oration of the Gods, 1872*: *None of these gods could give a true account of the creation of this little earth. All were woefully deficient in geology and astronomy. As a rule, they were most miserable legislators, and as executives, they were far inferior to the average of American presidents.*

Certainly, there have been a number of places in the Bible where God has showed himself to be less than astute. Why, for instance, would God inspire the author of Genesis 10:20 to claim various lands and nations had different languages (tongues) already, and then in Genesis 11:1 to claim *the whole earth was of one language?* And why would he *confuse their language, that they may not understand one another's speech,* in order to keep the people of Babel from building their tower? Did he really believe they might build it all the way to heaven: *Now nothing will be restrained from them, which they have imagined to do.* (Genesis 11:6) God could have just watched to see how far they got before they gave up instead of deciding to sow misperception and mistrust among the peoples of the world. And if God was bothered by the people getting too close to heaven then, is God bothered by all the manmade satellites in the 'heavens' now? Is getting closer to heaven all that bad?

But in the most mysterious way God the Father has revealed his almighty power in the voluntary humiliation and Resurrection of his Son, by which he conquered evil. (CCC 272)

An almighty God, it seems to me, should have been able to conquer evil by sacrificing a goat or rabbit or a rose bush. Or better yet, just going *Poof! Evil is conquered.*

Suicidal

Jesus plans to go to Jerusalem knowing he would be crucified and stay dead for three days.

Terroristic

I will bring terror on you . . . I will bring the sword against you . . . I will send pestilence among you, and you shall be delivered into enemy hands . . . I will devastate the land. (Leviticus 26:16-32)

You shall worship the Lord your God . . . I will send my terror in front of you, and will throw into confusion all the people against whom you shall come. (Exodus 23:27)

I will heap disasters upon them, spend my arrows against them, . . . In the street the sword shall bereave, and in the chambers terror, for young man and woman alike, nursing child and old gray head. (Deuteronomy 32:23–25)

. . . and the terror of God was upon the cities that were round about them. . . . (Genesis 35:5)

For thus says the Lord God: Bring up an assembly against them, and make them an object of terror and of plunder. The assembly shall stone them and with their swords they shall cut them down; they shall kill their sons and their daughters, and burn up their houses. (Ezekiel 23:46–47, FFRF)

Threatening and bullying

But of the tree of the knowledge of good and evil, thou shalt not eat of it: for in the day that thou eatest thereof thou shalt die. (Genesis 2:17)

The threat was an empty one, obviously over-the-top, since it would have ended humanity right there. And the devil promised something else:

And the serpent said unto the woman, Ye shall not surely die. (Genesis 3:4)

God lied and the devil told Eve the truth.

The truth that God is at work in all the actions of his creatures is inseparable from faith in God the Creator. (CCC 308)

God created the snake, knowing the consequence of that act. Who gave the snake the ability to talk in the language of Adam and Eve? It's like God was a co-conspirator.

Unforgiving

God is apparently reluctant to forgive. Instead he simply had his son killed on a cross and expected that to atone for the sinful nature he gave all humans. After God's feigned surprise following Adam and Eve's disobedience, he could have let them go hungry a few days for punishment and then forgiven them. Instead, he decided to punish them—and all of humanity—for all eternity.

Or he could have said *Oops! My bad*, let Adam and Eve die a *natural* death, and then created an Adam 2.0 and Eve 2.0 that were more inclined to be obedient to him. He had all eternity to keep trying until he created humans with the characteristics he desired.

Unmerciful

And I will dash them one against another, even the fathers and the sons together, saith the Lord: I will not pity, nor spare, nor have mercy, but destroy them. (Jeremiah 13:14)

A virgin must marry her rapist.

If a man find a damsel that is a virgin, who is not espoused, and taking her, lie with her, and the matter comes to judgment: He that lay with her shall give to the father of the maid fifty sicles of silver, and shall have her to wife, because he hath humbled her: he may not put her away all the days of her life. (Deuteronomy 22: 28-29)

Vindictive

The righteous shall rejoice when he sees the vengeance: he shall wash his feet in the blood of the wicked. (Psalm 58:10)

Say not, I will do so to him as he hath done to me: I will render to the man according to his work. (Proverbs 24:29)

Warmonger

The LORD is a man of war: the LORD is his name. (Exodus 15:3)

So Jephthah passed over unto the children of Ammon to fight against them; and the Lord delivered them into his hands. (Judges 11:32)

And when you win, save the virgins for yourselves:

Now kill all the boys. And kill every woman who has slept with a man, but save for yourselves every girl who has never slept with a man. (Numbers 31:17-18)

Then the Lord rained upon Sodom and upon Gomorrah brimstone and fire from the Lord out of heaven; And he overthrew those cities, and all the plain, and all the inhabitants of the cities, and that which grew upon the ground. (Judges 19:24-25)

Wimpy

Easily defeated. Forever, God had been destroying buildings with scary, destructive lightening strokes. Ben Franklin ends that run with the discovery that a conductive lightening rod would ground God's lightning game.

Wrathful

He that believeth on the Son hath everlasting life: and he that believeth not the Son shall not see life; but the wrath of God abideth on him. (John 3:36)

The faithful are also to be taught that the punishments here threatened await the third and fourth generation of the impious and wicked; not that the children are always chastised for the sins of their ancestors, but that while these and their children may go unpunished, their posterity shall not all escape the wrath and vengeance of the Almighty. (CCT)

God, says St. Augustine, *denies some things in His mercy which He grants in His wrath.* (CCT)

Do these attributes seem like the qualities you expect in a god/God or more like human basket-case qualities? Supposedly God created man in his image. What does it say about his image when the first two humans he created failed by sinning? Did he know it wouldn't be a problem since he would eventually *die by shedding blood* to save them? Does it mean sinning was part of his nature too? I'm left wondering what part of his image do any humans possess—and if there's any part of this creation story that makes any sense.

Actually things may be even worse than they were in Adam and Eve's day. Faith alone is not enough to be saved; you also need to follow the Ten Commandments and be baptized.

The Council of Trent [1569] teaches that the Ten Commandments are obligatory for Christians and that the justified [saved] man is still bound to keep them; the Second Vatican Council [1962] confirms: 'The bishops, successors of the apostles, receive from the Lord . . . the mission of teaching all peoples, and of preaching the Gospel to every creature, so that all men may attain salvation through faith, Baptism and the observance of the Commandments.' (CCC 2068)

The Ten (or More) Commandments

If ye love me, keep my commandments. (John 14:15)

What Christians and Jews traditionally recognize as the Ten Commandments is Exodus 20:2–17 and Deuteronomy 5:6–21. The commandments are also mentioned in Exodus 32, where they are not enumerated because they were written on stone tablets that were smashed by Moses; in Exodus 34, where they were written on replacement tablets; and in Deuteronomy 27, where they are referred to as a whole. None of these stone originals exist today.

Although the Bible uses the Hebrew phrase ***aseret had'varim*** (translated as *ten words*), the Ten Commandments in Exodus 20 and Deuteronomy 5 contain more than ten imperative statements, totaling fourteen or fifteen in all. Because the imperative statements are not specifically enumerated, various religions number the commandments differently. The following commandments are found in the Youth Catechism of the Catholic Church, (See YOUCAT 349) as espoused in the 5th century CE by St. Augustine. (See CCC 2066)

First Commandment:
I am the Lord, your God. You shall not have strange Gods before me.

Thus saith the Lord the King of Israel, and his redeemer the Lord of hosts; I am the first, and I am the last; and beside me there is no God. (Isaiah 44:6)

Now I know that the Lord is greater than all gods: for in the thing wherein they dealt proudly he was above them. (Exodus 18:11)

Thou shalt not have strange gods before me, it is equivalent to saying: Thou shalt worship me the true God; thou shalt not worship strange gods. (CCT)

So God thinks there are other gods and is threatened by the unauthorized *strange* ones? God needs to think about the fact that he thinks it is all the other gods that are strange. This reminds me of the mom looking out her window, watching the marching band go by, who exclaims, *My gosh! Everybody is out of step except my son.*

And thou shalt love the Lord thy God with all thy heart, and with all thy soul, and with all thy mind, and with all thy strength: this is the first commandment. (Mark 12:30)

The first commandment summons man to believe in God, to hope in him, and to love him above all else. (CCC 2134) Here you are commanded to love God. (You can command someone to love?)

So, what about atheists, people who don't believe there is evidence for any god?

Since it rejects or denies the existence of God, atheism is a sin against the first commandment. (CCC 2140)

Atheism must therefore be regarded as one of the most serious problems of our time. (CCC 2123)

Then again . . .

The imputability of this [atheism] offense can be significantly diminished in virtue of the intentions and the circumstances. . . . (CCC 2125) So atheism may be legitimate after all?

Nobody may be forced to act against his convictions, nor is anyone to be restrained from acting in accordance with his conscience in religious matters. . . . (CCC 2106)

Is atheism always a sin against the First Commandment? If someone intentionally denies God, he sins against the first commandment. However, his responsibility may be seriously impaired, as when, for example, he has learned nothing about God or has examined the question about God's existence conscientiously and cannot believe. The line between being unable to believe and being unwilling to believe is not clear. The attitude that simply dismisses faith as unimportant, without having examined it more closely, is often worse than well-considered ATHEISM. (YOUCAT 357)

And ye shall know the truth, and the truth shall make you free. (John 8:32)

Interestingly, the Catholic Church also claims: *This [First] commandment forbids us to: to acquire spiritual power through corruption. . . .* (YOUCAT 355) I hadn't realized the church knew that one could *acquire spiritual power through corruption*. How do they know that? Should we ask a Catholic bishop?

Nothing in the American legal system is designed to support this commandment. You are guaranteed the freedom to choose your own favorite god or ignore them all.

In *Why Are You Atheists So Angry?*, Greta Christina explains why belief matters: *I'm angry that when people run for political office in the United States, it's considered legitimate to grill them about their employment background, their positions on legislation, their positions on social issues, the taxes they've paid, even their sexual history . . . but it's considered invasive and intolerant to ask if they believe in talking snakes, demonic possession, magic underwear, magic crackers that turn into the flesh of their god, an earth that was created 6,000 years ago, or a god who put himself on Earth in human form and then sacrificed himself to himself to atone for sins that other people committed and to save humanity from the punishment he himself was planning to dole out. If someone is going to make decisions about science funding, emerging medical technology, our educational system, and so on . . . I think it matters if they believe any of that shit, and I bloody well want to know about it.*

Second Commandment:
You shall not take the name of the Lord your God in vain.

Blasphemy is directly opposed to the second commandment. . . . The prohibition of blasphemy extends to language against Christ's Church, the saints, and sacred things. . . . (CCC 2148)

This is called the *Genius Commandment* since it forbids any criticism of the church, which says you must obey these commandments in order to be saved. God gave these to Moses, which proves God was smarter than Moses. Without this commandment, the whole thing could unravel.

But he that shall blaspheme against the Holy Ghost hath never forgiveness, but is in danger of eternal damnation. (Mark 3:29)

And he that blasphemeth the name of the Lord, he shall surely be put to death, and all the congregation shall certainly stone him. . . . (Leviticus 24:16)

And whosoever shall speak a word against the Son of man, it shall be forgiven him: but unto him that blasphemeth against the Holy Ghost it shall not be forgiven. (Luke 12:10, Matthew 12:32)

So, it's forgivable to blaspheme Jesus Christ, but not God nor the Holy Ghost? According to Trinitarianism, are they not the same God? In any case, the US government and its constitution, representing American civil society, is already on the side of free speech, and not supporting religious-based speech restrictions.

Third Commandment:
Remember to keep holy the Lord's day.

Six days may work be done; but in the seventh is the sabbath of rest, holy to the Lord: whosoever doeth any work in the sabbath day, he shall surely be put to death. (Exodus 31:15)

Business enterprises, supported by their retail customers, broke from the Bible many decades ago. As evidence, you may note that Walmart, McDonald's, Target, Best Buy, and the Mall of America are open for business on Saturdays and Sundays.

Rulers and magistrates should be admonished and exhorted to lend the sanction and support of their authority to the pastors of the Church, particularly in upholding and extending the worship of God, and in commanding obedience to the injunctions of the priests. (CCT)

God's action is the model for human action. If God 'rested and was refreshed' on the seventh day, man too ought to 'rest.' (CCC 2172)

If *God's action is the model for human action* is really a true moral prerogative, then would it be moral for humans to destroy nearly every living thing on the earth, as God did in Noah's time?

Besides which—resting is required when you're omnipotent? God spending a whole day resting guts the idea of omnipotence to the core. An exhausted teenager or toddler needs at most only a few hours of rest and is ready to go again.

Fourth Commandment: Honor your father and your mother.

Even if they've beaten, abused, neglected, or abandoned you?

children are, of course, to prefer the will of God to the desires of their parents, always keeping in view the divine maxim: We ought to obey God rather than men. (CCT)

What if you tell your parents that you had a vision and God told you to do something? Are they required to believe you?

Honour thy father and thy mother: that thy days may be long upon the land which the Lord thy God giveth thee. (Exodus 20:12)

This is not a legal requirement in the US, where parental honor is expected and common, but earned.

The Fourth Commandment has many common sources outside and within the Bible, but is also contradicted by Luke, quoting God:

If any man come to me, and hate not his father, and mother, and wife, and children, and brethren, and sisters, yea, and his own life also, he cannot be my disciple. (Luke 14:26)

Besides our parents, the fourth commandment obliges us to respect and obey all our lawful superiors. (Baltimore Catechism) *Let every soul be subject unto the higher powers. For there is no power but of God: the powers that be are ordained of God. Whosoever therefore resisteth the power, resisteth the ordinance of God: and they that resist shall receive to themselves damnation.* (Romans 13:1-2)

This means, of course, that you must also honor and obey your priest.

Of those who resist the priest it is written: He that will be proud, and refuse to obey the commandment of the priest, who ministereth at that time to the Lord thy God, by the decree of the judge, that man shall die. (CCT)

Fifth Commandment:
You shall not kill.

This commandment appears simple and obvious and is accepted nearly everywhere.

The fifth commandment forbids direct and intentional killing as gravely sinful. (CCC 2268)

In the Sermon on the Mount, the Lord recalls the commandment, 'You shall not kill,' and adds to it the proscription of anger, hatred, and vengeance. (CCC 2262)

The Bible follows with:

Then said Jesus unto him, Put up again thy sword into his place: for all they that take the sword shall perish with the sword. (Matthew 26:52)

But upon closer examination of its foundation in the Bible, we find many exceptions to this *rule*. God has no trouble commanding the killing of everything in sight.

But of the cities of these people, which the Lord thy God doth give thee for an inheritance, thou shalt save alive nothing that breatheth: But thou shalt utterly destroy them; namely, the Hittites, and the Amorites, the Canaanites, and the Perizzites, the Hivites, and the Jebusites; as the Lord thy God hath commanded thee. (Deuteronomy 20:16-17)

As ordered by God, one should also intentionally kill someone that:

a) is gay (Leviticus 20:13)
b) is a witch (Exodus 22:18)
c) is unfaithful (Leviticus 20:10)
d) hits their father (Exodus 21:15)
e) is disobedient (Deuteronomy 21:18-21)
f) is not a virgin (Deuteronomy 22:20-21)
g) plays the whore (Leviticus 21:9)
h) is a fortuneteller (Leviticus 20:27)
i) is a non-believer (2 Chronicles 15:12-13)
j) works on the Sabbath (Exodus 31:12-15)
k) curses their parents (Leviticus 20:9, Exodus 21:17)
l) is the son of a sinner (Isaiah 14:21)
m) blasphemes the Lord (Leviticus 24:11-16)
n) is a child of your enemy (Numbers 31:17)
o) makes fun of a bald person (2 Kings 2:23-24)
p) worships deities of sun, moon, or stars (Deuteronomy 17:3-5)
q) is tempted by family to worship other gods (Deuteronomy 13:6-9)
r) is an engaged virgin that is raped and doesn't cry out (Deuteronomy 22:23-24)

If your god asks you to kill his enemies, that's okay.

But those mine enemies, which would not that I should reign over them, bring hither, and slay them before me. (Luke 19:27)

This leads one today, if you accept God as an example, to be tempted to think God's enemies should be killed. The morality of this has evolved over the last 2,000 years, enabling us to say, *No.*

Killing by accident or in self-defense is accepted as moral. Killing intentionally is not moral, usually. In the words of St. Augustine:

God forbid that what we do for a good and lawful end shall be imputed to us, if, contrary to our intention, evil thereby befall any one. (CCT)

So it appears that what you are thinking as you kill someone determines if the killing is accepted as moral. Recall, however, this admonition: ***The end does not justify the means.*** (YOUCAT 292) Morality here depends on the situation. Killing in self-defense is always considered moral—which means that the end, your survival, does justify the means, the killing.

Faith gives us the certainty that God would not permit an evil if he did not cause a good to come from that very evil, by ways that we shall fully know only in eternal life. (CCC 324)
A true mystery.

We also tend to accept our wars as moral because they are purportedly fought in our self-defense. But opponents' wars are considered evil. Even collateral damage such as the killing of babies and children in the bombings of Dresden and Hiroshima is accepted by most as a legitimate cost of war. How do we determine that? In Noah's time, (Genesis 6-9) God flooded the whole world, presumably drowning many innocent babies and children. And not long after that, he destroyed Sodom and Gomorrah. (See Genesis 13:10 and Genesis 18-19) The concept of acceptable collateral damage goes back a long way.

Recall in Part II the section on morality where it's advocated that morality begins and ends with the people. At least it does not originate with God or the church. It is interesting that in an area that needs good moral judgment the most, going to war, which involves lots of killing, the church declares that morality is the responsibility of the people making the war. After listing four conditions for *legitimate defense by military force,* the church continues: *The evaluation of these conditions for moral legitimacy belongs to the prudential judgment of those who have responsibility for the common good.* (CCC 2309) Thus, it is moral for the people's representatives to decide the morality of war and presumably God would accept their judgment.

Is a war launched in order to prevent an attack ever moral and just? What justifies a *preemptive* war is a hotly debated issue today. Apparently God is leaving it to us mortals to answer that question.

Could it be that even God changes His mind regarding killing? In the era of the Inquisition the Catholic Church had faith that God approved of their killing of non-believers. Some time after that mean streak we took away the right of the church to kill anybody. The Catholic Church has since evolved to become an opponent of the death penalty. I assume Catholics have faith that God supports this new stance.

Assuming that the guilty party's identity and responsibility have been fully determined, the traditional teaching of the Church does not exclude recourse to the death penalty, if this is the only possible way of effectively defending human lives against the unjust aggressor. (CCC 2267)

The notion that it's fair and just to kill someone who killed another is still evolving. Some Christians and civil governments accept the court ordered death penalty. Some don't. The number of states abolishing the death penalty is over 18 and growing. Note that the voters are deciding, not God. It seems that any court that endorses a death penalty and desires to display the Ten Commandments on its premises should also want to add this display: *The judges and juries herein reject the Fifth Commandment.*

Suicide is seriously contrary to justice, hope, and charity. It is forbidden by the fifth commandment. (CCC 2325)

How is Jesus voluntarily walking into Jerusalem that fateful day any different from you deliberately stepping in front of a speeding train today? Christianity is built on an assisted suicide story. Suicide was acceptable for Jesus but it's forbidden for you? The Catholic Church seems to be saying that an eleven-year-old student who is bullied so severely that the resulting desperation induces depression which leads to suicide is also violating the Fifth

Commandment and is thus destined for eternal hell. Doesn't this make the church also look like the ultimate bully? There is something wrong here. This is evil.

The Catholic Church formerly practiced that persons committing suicide could not have a Catholic funeral mass and burial. However, the Church has since changed this practice. (Wikipedia)
Evolution, yes?

We should not despair of the eternal salvation of persons who have taken their own lives. By ways known to him alone, God can provide the opportunity for salutary repentance. The Church prays for persons who have taken their own lives. (CCC 2283)

Does the church also pray for those who violate other commandments?

It is worth noting that a human sacrifice and a human killing are both murder in this setting. If today you would sacrifice your only son by fire, gunshot, or crucifixion, you would be hauled in quickly and charged with murder. You might claim you did it to atone for something others did long ago or something others will do in the future, but that is not self-defense. Your son is innocent. You are guilty.

Is everlasting punishment ever morally okay?

And these shall go away into everlasting punishment: but the righteous into life eternal. (Matthew 25:46)

The teaching of the Church affirms the existence of hell and its eternity. (CCC 1035)

You must ask: Is torturing someone forever an acceptable form of justice? Isn't it a cruel and unusual punishment? If it is, the 8th Amendment to the US Constitution forbids it. Thus, the framers of the Constitution broke with the Christian Bible more than two centuries ago. Maybe it's no accident that neither God nor the Bible is ever mentioned, not even once, in the Constitution of the United States.

Interestingly, a Pew Research Center's Forum survey of five years ago and a recent (2014) *Washington Post*/ABC News poll show that Christians are more likely to support torture. The non-religious are less likely to do so.

As to the conflict between the Constitution and Christianity, fixing it is simple, says Baptist preacher and sometime-presidential candidate Mike Huckabee (2008): *Change the Constitution. But I believe it's a lot easier to change the Constitution than it would be to change the word of the living God. And that's what we need to do—to amend the Constitution so it's in God's standards rather than try to change God's standards so it lines up with some contemporary view.* Mike, over two hundred years ago our brilliant founding fathers decided to break with God's standard. You want to take us back more than two centuries? I'm curious: What would you suggest we drop or add in the constitution? Be clear.

Sixth Commandment:
You shall not commit adultery.

And the man that committeth adultery with another man's wife, even he that committeth adultery with his neighbour's wife, the adulterer and the adulteress shall surely be put to death. (Leviticus 20:10)

Sounds reasonable to everyone, including many scriptwriters, but not enough to make it illegal, at least not in the United States. If the state should support the Ten Commandments, shouldn't it also make violations of them illegal?

It gets worse: *But I say unto you, That whosoever looketh on a woman to lust after her hath committed adultery with her already in his heart.* (Matthew 5:28) A quaint idea, but we have not punished thought crimes for centuries.

Seventh Commandment:
You shall not steal.

This commandment is pretty much accepted everywhere. It is not unique to nor owned by Christians. It even applies to stealing people:

To enslave a freeman, or appropriate the slave of another is called man-stealing. (CCT) On the other hand, it doesn't prevent you from owning people that others might want to steal.

Eighth Commandment:
You shall not bear false witness against your neighbor.

But it's okay to bear false witness against others?

This idea is not unique to Judaism or Christianity. It applies to more than everyday life and/or judicial proceedings of various sorts too: It's the commandment behind the warning to Catholic bishops and priests that they are not to propagate unsound Catholic doctrine:

But of all sorts of calumnies the worst is that which is directed against Catholic doctrine and its teachers. Persons who extol the propagators of error and of unsound doctrine are guilty of a like crime. (CCT)

Ninth Commandment:
You shall not covet your neighbor's wife.

Your neighbor's wife is her husband's possession, after all.

The Ninth Commandment forbids, not desire per se, but rather disordered desires. . . . The erotic attraction between man and woman was created by God and is therefore good; it is part of a person's sexual nature and biological constitution. (YOUCAT 462)

Still, as noted above, we don't consider thoughts to be crimes anymore. That died with the Dark Ages.

Tenth Commandment:
You shall not covet your neighbor's goods.

Not his ox, or his ass, *or anything that is your neighbor's*. This is definitely at odds with most contemporary capitalistic advertising and marketing strategies, which are rooted in encouraging consumers to *keep up with* the neighbor next door or even do them one—or two—better.

A Christian must learn to distinguish reasonable desires from those that are unreasonable and adjust and to acquire an interior attitude of respect for other people's property. (YOUCAT 465)

The dividing line between reasonable desire and unreasonable/disordered desire in the last two commandments is admittedly not at all well-defined. Christians have to learn for themselves, the hard way. The problem is that God decides if you judged as he does. And he won't tell you his decision ahead of time. If you ever-so-slightly miss, it's to hell for you. Is that clear? Is that fair?

For whosoever shall keep the whole law, and yet offend in one point, he is guilty of all. (James 2:10)
You must follow all these commandments perfectly or go to hell.

The Supreme Court on July 27, 2005, in the case of McCreary County v. ACLU ruled that a display of the Ten Commandments on government property was unconstitutional. The US government cannot enforce contradictory and controversial Christian commandments. The US government cannot resolve variations in commandments among different sects. There are still no commandments regarding: war, genocide, child abuse, starvation, slavery, human trafficking, bullying, income redistribution, terrorism, overpopulation, etc.

Redistribution of Wealth

Speaking of income redistribution—this is not one of the commandments, but a topic that comes up regularly in the New Testament, in particular. An important precept of Christian ethics, albeit one that is too frequently ignored in the opinion of those interested in social justice, regards the relationship between money and morality or redemption in the Bible.

It is easier for a camel to go through the eye of a needle than for a rich man to enter the kingdom of God. (Mark 10:25)

For it is easier for a camel to go through the eye of a needle than for a rich man to enter the kingdom of God. (Luke 18:25)

God really wants us to pay attention to this.

Again I say to you, it is easier for a camel to go through the eye of a needle, than for a rich man to enter the kingdom of God. (Matthew 19:24)

Third time's the charm?

But whoso hath this world's good, and seeth his brother have need, and shutteth up his bowels of compassion from him, how dwelleth the love of God in him? (1 John 3:17)

What should the rich man do with his wealth, then?

Let the man with two tunics share with him who has none, and let him who has food do likewise. (Luke 3:11)

When the Son of man shall come in his glory, and all the holy angels with him, then shall he sit upon the throne of his glory: And before him shall be gathered all nations: and he shall separate them one from another, as a shepherd divideth his sheep from the goats: And he shall set the sheep on his right hand, but the goats on the left. Then shall the King say unto them on his right hand, Come, ye blessed of my Father, inherit the kingdom prepared for you from the foundation of the world: For I was an hungred, and ye gave me meat: I was thirsty, and ye gave me drink: I was a stranger, and ye took me in: Naked, and ye clothed me: I was sick, and ye visited me: I was in prison, and ye came unto me. Then shall the righteous answer him, saying, Lord, when saw we thee an hungred, and fed thee? or thirsty, and gave thee drink? When saw we thee a stranger, and took thee in? or naked, and clothed thee? Or when saw we thee sick, or in prison, and came unto thee?

And the King shall answer and say unto them, Verily I say unto you, Inasmuch as ye have done it unto one of the least of these my brethren, ye have done it unto me. Then shall he say also unto them on the left hand, Depart from me, ye cursed, into everlasting fire, prepared for the devil and his angels: For I was an hungred, and ye gave me no meat: I was thirsty, and ye gave me no drink: I was a stranger, and ye took me not in: naked, and ye clothed me not: sick, and in prison, and ye visited me not. Then shall they also answer him, saying, Lord, when saw we thee an hungred, or athirst, or a stranger, or naked, or sick, or in prison, and did not minister unto thee? Then shall he answer them, saying, Verily I say unto you, Inasmuch as ye did it not to one of the least of these, ye did it not to me. And these shall go away into everlasting punishment: but the righteous into life eternal. (Matthew 25:31-46)

Go to now, ye rich men, weep and howl for your miseries that shall come upon you. Your riches are corrupted, and your garments are motheaten. Your gold and silver is cankered; and the rust of them shall be a witness against you, and shall eat your flesh as it were fire. Ye have heaped treasure together for the last days. Behold, the hire of the labourers who have reaped down your fields, which is of you kept back by fraud, crieth: and the cries of them which have reaped are entered into the ears of the Lord of sabaoth. Ye have lived in pleasure on the earth, and been wanton; ye have nourished your hearts, as in a day of slaughter. Ye have condemned and killed the just; and he doth not resist you. (James 5:1-6)

No one can serve two masters; for either he will hate the one and love the other, or else he will hold to one and despise the other. You cannot serve God and Money. (Matthew 6:24)

The early Christians clearly demonstrated a communistic/socialistic view:

Neither was there any among them that lacked: for as many as were possessors of lands or houses sold them, and brought the prices of the things that were sold, And laid them down at the apostles' feet: and distribution was made unto every man according as he had need. And Joses, who by the apostles was surnamed Barnabas (which is, being interpreted, the son of consolation), a Levite, and of the country of Cyprus, Having land, sold it, and brought the money, and laid it at the apostles' feet. (Acts 4:34-37)

And all that believed were together, and had all things common; And sold their possessions and goods, and parted them to all men, as every man had need. (Acts 2:44-45)

For even when we were with you, this we commanded you, that if any would not work, neither should he eat. (2 Thessalonians 3:10)

Here eating is tied to work, but no connection is made with whether work is available or if a handicap does not allow work.

He that is greedy of gain troubleth his own house; but he that hateth gifts shall live. (Proverbs 15:26-27)

And until a more equitable distribution of wealth is achieved, the ills of our society will not be resolved. . . . 'Trickle down' theories have been shown to be erroneous: it is not enough to hope that the poor may gather the crumbs that fall from the table of the rich. It is necessary to take direct action in favour of the most disadvantaged, attention to should be a priority for governors, as it is for the smallest within a family. (Pope Francis, VIS April 11, 2015)

Fact or Fiction?

Adults accept trappings of Christ—cathedrals, vestments, bible, priests and hosts—as the real thing without realizing that it all may be a fiction.

Whence do we chiefly derive our historical knowledge of Jesus Christ, His life and teachings, and of the Church He established? We derive our historical knowledge of Jesus Christ, His life and teachings, and of the Church He established chiefly from the books of the Bible which can be proved to be reliable historical records. (Baltimore Catechism, 1941)

Robert G. Ingersoll, over a century ago, had this to say about the Christian story: *The notion that faith in Christ is to be rewarded by an eternity of bliss, while a dependence upon reason, observation, and experience merits everlasting pain, is too absurd for refutation, and can be believed only by that unhappy mixture of insanity and ignorance called faith.* (D. M. Murdock, author of *Who Was Jesus?*)

Christians are followers of Christ and believe in the inspired words of God, the Bible. And how do we know we can trust that the Bible is the inspired word of God? The Bible says so. *All Scripture is given by inspiration of God, and is profitable for doctrine, for reproof, for correction, for instruction in righteousness:* (2 Timothy 3:16) This is the classic biblical circular argument.

If God could inspire people to write something special, and apparently many writers were involved, why couldn't God have taken the shortcut and just inspired us all to be good in the first place and thus saved?

Next we will examine, even more closely, a Christian Bible.

Part IV

The Bible

Properly read, the Bible is the most potent force for atheism ever conceived. (Isaac Asimov)

Many religions have sacred scriptures, e.g., the Talmud, the Qur'an, texts which are central to their religious tradition and practices and which they consider to be divinely or supernaturally revealed or inspired. The Bible is the holy book of Christianity.

Repent ye, and believe the gospel. (Mark 1:15)

It is not surprising that something that was codified 17 centuries ago, cobbled together from accounts dating back a millennium before that, many grounded in oral traditions and expressed in multiple languages, recorded and copied by scribes of varying degrees of expertise, and then translated more or less accurately into other languages, is filled not only with material at odds with modern science, but also with verses that directly contradict each other (sometimes even within the same book or chapter). What is surprising is the number of people who view every word in this mass of contradictions, ambiguities, and contrivances as *inerrant*—that is, incapable of being wrong; who declare that the Bible (of which there are scores of different versions in English alone) is without defect.

How do they know that? The Bible says so.

All scripture is given by inspiration of God, and is profitable for doctrine, for reproof, for correction, for instruction in righteousness. (2 Timothy 3:16)

The Sacred Scriptures contain the Word of God and, because they are inspired, they are truly the Word of God. (CCC 135)

Every word of God is pure. (Proverbs 30:5)

God is the author of Sacred Scripture because he inspired its human authors; he acts in them and by means of them. He thus gives assurance that their writings teach without error his saving truth. (CCC 136)

[The Bible] is therefore inerrant in the originals and authoritative in all matters. . . . (Doctrinal Statement, August 3, 2014, Liberty University)

The original languages of all the books of the Old and New Testaments are not known nor where they were written nor by whom. For that matter, neither is the day and year of Jesus's birth or of his execution.

We do not possess the original gospels, and it is an indisputable fact that even the most ancient copies of the New Testament have been worked over repeatedly by a number of hands and do not agree with each other in many places, making the task of determining what was in the original extremely difficult, if not impossible. (D. M. Murdock, author of *Who Was Jesus?*)

We don't have the originals! We have only error-ridden copies . . . (Dr. Bart Ehrman, author of *Misquoting Jesus*)

No television preacher has ever read the Bible. Neither has any evangelical politician. Neither has the pope. Neither have I. And neither have you. At best, we've all read a bad translation—translation of translations of translations of hand-copied copies of copies of copies of copies, and on and on, hundreds of times. (Kurt Eichenwald, *Newsweek*, December 23, 2014)

Is a Bible written centuries ago still valid and true today? The Bible says the earth is immovable and flat and the sky is solid, and it endorses genocide by God, as well as slavery. Even so, Gallup reported in 2014 that three out of four people in the US still view the Bible as the *word of God.*

But the word of the Lord endureth for ever. And this is the word which by the gospel is preached unto you. (1 Peter 1:25)

Let's look at the foundation of the Bible, its early history.

Early Bible stories were told verbally and in a variety of languages. Jesus and the disciples would likely have spoken Aramaic. The Dead Sea Scrolls, the second-oldest known surviving manuscripts of works later included in the Hebrew Bible canon, are written in Hebrew, with some in Aramaic (in different regional dialects), and a few in

Greek. They date from the last three centuries BCE and the 1st century CE.

At that time the literacy rate was about 10 percent. While Paul may have been literate, his letters were written by scribes, not Paul, and in Greek. (Besides, Paul never met Jesus even though they were contemporaries. At least he never mentioned doing so in his letters, and that doesn't seem like something he would have neglected to comment on.)

The first official New Testament Canon of Sacred Scriptures (the New Testament of the Bible) and the Nicene Creed, both still used today, were created in 325 CE when Roman Emperor Constantine called together the Council of Nicaea to settle infighting among various powerful Christian bishops that threatened to contribute to discord and potential splits in the Roman empire. The Roman Council in 382 CE along with the Council of Hippo in 393 CE and the Council of Carthage in 397 CE further firmed up the New Testament canon that is generally accepted today.

Which of the many original Christian writings are really inspired by the Holy Spirit has been defined since the fourth century in the so-called Canon of Sacred Scriptures. (YOUCAT 14)

As recently as the Second Vatican Council, held in 1962–1965, the church was still reinforcing this precept:

The books of Scripture firmly, faithfully, and without error teach the truth. . . . Written under the inspiration of the Holy Spirit, they have God as their author. (CCC 105)

Revelation, the last book of the New Testament, was officially adopted early in the 5th century CE. That wasn't the last time the Bible was brought up to date, however.

In 480 CE St. Jerome translated the Bible from Late Latin (and a few bits in Hebrew, Aramaic, and Greek) to Latin. Called the Latin Vulgate Bible, this Bible was declared in 1546 to be the only legitimate Catholic Church version.

In 1370 John Wycliff promoted the first English vernacular (Middle

English) Bible, translated from Latin, which soon was declared illegal to possess.

Martin Luther published his German translation of the New Testament in 1522, and he and his collaborators completed the translation of the Old Testament in 1534. Others had translated the Bible into German, but critics noted that Luther was prone to tailoring his translation to his own doctrine.

In 1526, William Tyndale published the New Testament of the Bible, in which he drew directly from Hebrew and Greek texts. In 1536, he was convicted of heresy and executed by strangulation, after which his body was burnt at the stake.

In 1604, the King James Bible was commissioned. It was published in 1611. Seventy percent of it came from the Tyndall Bible. It became a best seller.

We know today that:

None of the Old Testament writers were eyewitness to the events they wrote about.

No New Testament authors ever met Jesus.

No one knows who the original authors of the Gospels were.

We have no original documents written by Jesus.

We have no original documents written by anyone who met Jesus.

No secular contemporary authors in the 1st century wrote anything referring to Jesus.

Every word in the Bible is therefore hearsay, i.e., nothing that would be accepted in a court of law in the United States. Nor is it science, even though Pope Francis, speaking to the United Bible Societies in Italy in 2014, said:

Yours is the fruit of a patient, careful, fraternal, competent and, above all, faithful work. If you do not believe, you do not understand; if you do not believe, you cannot stand firm . . . I would very much like all Christians to be able to learn 'the sublime science of Jesus Christ' through frequent reading of the Word of God. . . . (Vatican City, September 29, 2014)

As someone who trained to be a chemical technician, the pope should be aware that science relies upon actual evidence, not merely faith.

In the Beginning Were the Words . . .

Before the invention of the printing press, producing a Bible was the work of scribes, monks, and virtually anyone that wanted the laborious task of hand-copying a previous version. There was not even a requirement that one had to be able to read to do that. Mistakes were inevitable.

And, before they were copied, these works had to be translated, of course. We know very little of the qualifications of those who did this work. As noted, early on text in Hebrew and Aramaic was common. These eventually evolved into Greek and then Latin translations. Translations into the modern vernacular appeared soon after the invention of the printing press.

That means that the Bible you're reading today likely went through at least four language translations, with the first copies being written-down versions of various oral traditions. If you want to get an interesting picture of what that might result in, simulate it with your computer. Take a passage in English and, using Google or Bing, translate it to another language, then another, and then back to English. This would be akin to converting something in Aramaic to Greek, then to Latin and finally to English. Here's what happens when you translate the first paragraph of this section three times:

Original Text:
Before the invention of the printing press, producing a Bible was the work of scribes, monks, and virtually anyone that wanted the laborious task of hand-copying a previous version. There was not even a requirement that one had to be able to read to do that. Mistakes were inevitable.

Original Translated to Greek:
Πριν από την εφεύρεση της τυπογραφίας, που παράγουν ένα Βίβλος ήταν το έργο του γραφείς, οι μοναχοί, και σχεδόν ο

καθένας που ήθελε το επίπονο έργο του χεριού αντιγράφοντας μια προηγούμενη έκδοση. Δεν υπήρχε καν η υποχρέωση που έπρεπε να είναι σε θέση να διαβάσει για να το κάνουμε αυτό. Τα λάθη ήταν αναπόφευκτη.

Greek Translated to Latin:
Ante typographicam inventam producentibus Bible erat factura scribis monachi, et fere omnium qui volebant de labore manuum exscribend testatem version. Fuit etiam officium legere debebat facere. Erroribus subeunda essent.

Latin Translated to English:
Before the Bible was the work of teachers of the law is one that produces typographicam the discovery of the monks, and of almost all the labor of the hands of the man that would copying version. It was also the duty to read it should have to do. Mistakes were inevitable.

If you don't like what the computer does to your text, have three well-qualified humans do the three translations, independent of one another. And then imagine you must accept the final translation as an official inspired version.

There is no single *official* Bible, except maybe *The Beekeeper's Bible* or *The Golfer's Bible*. After nearly two thousand years we should not be surprised to see many variations. There are more than a thousand versions in circulation today, with nearly that many branches of Christianity claiming their own unique, true version. (The State Senate in Tennessee recently voted to make the Bible the State Book. (April 16, 2015) My question: which one?) While it is instructive to compare how they differ, it would be burdensome to compare them all. So let's simply examine some verses from a single chapter (Leviticus 19) of two Bibles, both of which have *Catholic* printed on their covers.

Version **A,** billed as the **Catholic** Public Domain Version, Sacred Bible, 2009 (derived from Latin Vulgate Bible & Challoner Douay Rheims Bible)

Version **B,** billed as the New Revised Standard Version, **Catholic** Edition (authorized by the National Conference of Catholic Bishops in the U.S.A.)

Leviticus 19:1

A: *The Lord spoke to Moses, saying . . .*

B: *And the Lord spoke to Moses . . .*

The passages are similar, but neither gives any hint of who the divinely inspired third-person writer was. A later storyteller?

Leviticus 19:3

A: *Let each one fear his father and his mother . . .*

B: *You shall each revere your mother and his father . . .*

Fear and *revere* are sometimes considered synonyms, but they have very different nuances.

Leviticus 19:5

A: If *you immolate a* victim *of peace offerings to the Lord . . .*

B: When *you offer a* sacrifice *of well-being to the Lord . . .*

If and *when* are two significantly different words. As are *victim* and *sacrifice.*

Leviticus 19:8

A: . . . *And that* soul *shall* perish *from his people.*

B: . . . *and any such* person *shall be* cut off *from the people.*

To a Catholic theologian, a *soul* and a *person* are definitely not the same thing. To *perish* is to die. To be *cut off* is to be separated from.

Leviticus 19:9

A: . . . *you shall not cut it down to the surface of the land . . .*

B: . . . *you shall not reap to the very edges of your field . . .*

Easy. And as a bonus, in the King James Version, the passage avoids the *corners.*

Leviticus 19:17

A: *You shall not hate your brother in your heart, b*ut reprove him openly . . .

B: *You shall not hate in your heart anyone of your kin . . .*

Brother and *kin* are not the same. The *reprove* phrase is missing altogether in Version B. Does this also mean you can hate people who are unrelated to you?

This is only a short sample, but you get the idea. Even within the same denomination, there are variations. Nevertheless, as Pope Benedict XV reminded us: The Bible is immune to error.

St. Jerome's teaching on this point serves to confirm and illustrate what our predecessor of happy memory, Leo XIII, declared to be the ancient and traditional belief of the Church touching the absolute immunity of Scripture from error: So far is it from being the case that error can be compatible with inspiration, that, on the contrary, it not only of its very nature precludes the presence of error, but as necessarily excludes it and forbids it as God, the Supreme Truth, necessarily cannot be the Author of error. (Spiritus Paraclitus Encyclical of Pope Benedict XV, September 15, 1920)

How can contradictory versions be free of error? Is this another one of those *mysteries*? The differences noted above are all relatively small potatoes that can be chalked up to nuances in translating. Let's take a closer look at some of the really big conflicts, beginning at the very beginning, looking at them with a scientific eye rather than through the eyes of faith.

The Old Testament

All early source documents are lost to history.

The author of Genesis, the Judeo-Christian origin myth, is unknown. It appears to be an early monotheistic creation story.

We can only judge what Genesis was all about based on evidence we have today about how the earth was formed.

The Creation Story

Genesis 1 offers the first version of man's creation. (The very next chapter, which we'll also look at, offers another, slightly different one.)

In the beginning . . .

Day 1—God created sky, earth, and light and separated it from darkness. No source is mentioned for the light.

Day 2—God separated the sky and waters. Early myths had water in ocean basins and beyond the sky, since rain brings water from the sky.

Day 3—God created land and gathered the seas and also created vegetation. Plants need sunlight.

And God said, Let the earth bring forth grass, the herb yielding seed, and the fruit tree yielding fruit after his kind, whose seed is in itself, upon the earth: and it was so. And the earth brought forth grass, and herb yielding seed after his kind, and the tree yielding fruit, whose seed was in itself, after his kind: and God saw that it was good. (Genesis 1:11-12)

That's a bit repetitive, but, as God might have mused, *Inspiring is easy. Editing is hard.*

Day 4—God created the sun, moon, and stars—to light the earth and cause day and night.

Day 5—God created sea life, birds, reptiles, and insects.

Day 6—God created Adam and Eve (at the same time).

And God said, Let the earth bring forth the living creature after his kind, cattle, and creeping thing, and beast of the earth after his kind: and it was so. And God made the beast of the earth after his kind, and cattle after their kind, and every thing that creepeth upon the earth after his kind: and God saw that it was good. (Genesis 1:24-25)

And God said, Let us make man in our image, after our likeness: and let them have dominion over the fish of the sea, and over the fowl of the air, and over the cattle, and over all the earth, and over every creeping thing that creepeth upon the earth. So God created man in his own image, in the image of God created he him; male and female created he them. And God blessed them, and God said unto them, Be fruitful, and multiply, and replenish the earth, and subdue it: and have dominion over the fish of the sea, and over the fowl of the air, and over every living thing that moveth upon the earth. (Genesis 1:26-28)

Day 7—God rested and made the day holy. (This resting stuff makes omnipotence look a bit puny, doesn't it?) Note that days, evenings, and *mornings* are mentioned long before the sun is created. There have been many conjectures among people of faith regarding the length of these days. None fit the scientific evidence.

The beginning of Verse 26 prompts other questions as well:

And God said, Let us make man in our image, after our likeness . . . (Genesis 1:26)

God *said* all this to whom? *Us*? Unless God is using the majestic plural form of address—and he could be doing that, being God and all—this could indicate someone was with God. Who? A God helper? Another god? Ditto: *our likeness*. And finally, who passed this story along orally to the first writer to record it?

Genesis 2, the second version of man's creation, might just be intended to reinforce the message of the first one. But . . . the accounts don't quite match.

Genesis 2:7—God created Adam [again] from the desolate earth.

Genesis 2:8—God places man in the Garden of Eden.

Genesis 2:15—God [again] places man in Garden of Eden.

Genesis 2:19—God [again] creates the beasts of the field and birds of the air.

Genesis 2:22—God [again] creates Eve from a rib of Adam.

Were humans created before or after the animals? Were Adam and Eve created at the same time or separately?

Version 1: *And God made the beast of the earth after his kind, and cattle after their kind, and every thing that creepeth upon the earth after his kind: and God saw that it was good. And God said, Let us make man in our image, after our likeness: and let them have dominion over the fish of the sea, and over the fowl of the air, and over the cattle, and over all the earth, and over every creeping thing that creepeth upon the earth. So God created man in his own image, in the image of God created he him; male and female created he them.* (Genesis 1:25-27)

Version 2: *And the Lord God said, It is not good that the man should be alone; I will make him an help mate for him. And out of the ground the Lord God formed every beast of the field, and every fowl of the air; and brought them unto Adam to see what he would call them: and whatsoever Adam called every living creature, that was the name thereof.* (Genesis 2:18-19)

How orderly was the process by which things were created?

1: Things are done step-by-step. The only discrepancy is that there is no sun or moon or stars on the first three *days*.

2: God fixes things up as he goes. (That's a funny thing for a being that is omniscient to have to do, isn't it?) He creates a garden, puts a man in it, and tells him he can eat from every tree except the tree of knowledge of good and evil. The man is lonely and not satisfied with animals as helpers. God creates a woman for him.

How satisfied with creation was he?

1: Evidently God was very satisfied. He says *it was good* after each of his labors and rests on the seventh day, happy with the way everything turned out.

2: God has to fix up his creation as he goes. And there at the end of Genesis 2, when the man and his wife were both naked and ashamed, he was definitely *not* happy with that!

So, which do you believe: Genesis 1 or Genesis 2? Neither, maybe? The two chapters obviously represent the clumsy merging of two different origin stories.

Adam and Eve

Both the Bible (Judaism and Christianity) and the Qur'an (Islam) affirm the Adam and Eve story.

And the Lord God commanded the man, saying, Of every tree of the garden thou mayest freely eat: But of the tree of the knowledge of good and evil, thou shalt not eat of it: for in the day that thou eatest thereof thou shalt surely die. (Genesis 2:16-17)

After God created Adam and Eve and before they sinned, he told them to be fruitful and multiply. (With whom? Can you think of any option that doesn't involve incest?) He did this despite knowing that a devastating flood lay in store for humanity down the road.

And God blessed them, and God said unto them, Be fruitful, and multiply, and replenish the earth, and subdue it. . . . (Genesis 1:28)

God, who creates and conserves all things by his Word, provides men with constant evidence of himself in created realities. And furthermore, wishing to open up the way to heavenly salvation—he manifested himself to our first parents from the very beginning. He invited them to intimate communion with himself and clothed them with resplendent grace and justice. (CCC 54)

Beyond the witness to himself that God gives in created things, he manifested himself to our first parents, spoke to them and, after the fall, promised them salvation and offered them his covenant. (CCC 70)

He also gave them Cain, son of Adam and Eve, who killed his brother and married his sister (or was it his mom, Eve?).

And Cain knew his wife; and she conceived, and bare Enoch: and he builded a city, and called the name of the city, after the name of his son, Enoch. (Genesis 4:17)

And thus the first family begins sinful humanity.

For as by one man's disobedience, many were made sinners. . . . (Romans 5:19)

Yes, Adam and Eve disobeyed and ate the apple, but it was God who defined that as a sin, determined the consequences of a sin and its effect on humanity, and established what it took to forgive a sin. But why was all this necessary? If God was actually omnibenevolent, then why not create all humans as he did in the case of his mother Mary, via an immaculate conception and thus free from original sin? That's the way I would have done it: assume everyone is good rather than the default bad we now have.

According to Catholic dogma, he could have done it. He did so for the Virgin Mary, who was conceived in her mother's womb free from original sin by virtue of the *foreseen merits* of her son Jesus. He had the power and authority. After all, his son did.

And Jesus came and spake unto them, saying, All power is given unto me in heaven and in earth. (Matthew 28:18)

We die because of sin.

Therefore, just as through one man sin entered into this world, and through sin, death; so also death was transferred to all men, to all who have sinned. (Romans 5:12)

What about all the other creatures that die? Did they also sin? Do they need sin to cause their deaths?

Maybe the best thing any human ever did was to disobey someone who claimed that knowledge of anything (especially good and evil) was immoral, evil, or to be avoided. And furthermore, if Adam and Eve were *created in the image and likeness of God* and God did not have to obey anyone, then it would be contradictory to think that *they* had to obey anyone. It was unfair, perhaps even evil, for God to expect them to be obedient.

Suppose, as a developer, you went to the Planning and Zoning Department and proposed a large new fantastic housing subdivision—but without parks, playgrounds and open spaces. The only place anyone could play was on the right-of-way grass in the middle of a four-lane street. Without fences. How far would you get if you suggested that people have the free will to choose to play on the grass or on the dangerous street? As a final plea in the application you also propose that you already know that a majority of people in the project will die playing in traffic. It's their fault. They chose to play in the street instead of staying on the grass. Is this rational? Would this get approval?

Back to Adam and Eve . . .

The Church teaches that the one true God, our Creator and Lord, can be known with certainty from his works, by the natural light of human reason. (CCC 47)

Does that include Adam and Eve?

We really can name God, starting from the manifold perfections of his creatures, which are likenesses of the infinitely perfect God, even if our limited language cannot exhaust the mystery. (CCC 48)

Does this include the *manifold perfections* of Adam and Eve? Couldn't God have created Adam and Eve so that obedience to God was a natural force, like gravity? If not, then he knew he was creating people that he would later have to try (and fail) to save from going to the hell he created, for violating the rules he created.

The human mind, in its turn, is hampered in the attaining of such truths, not only by the impact of the senses and the imagination, but also by disordered appetites which are the consequences of original sin. (CCC 37)

Appetites which are natural to *creatures, which are likenesses of the infinitely perfect God?*

The Church, which has the mind of Christ, knows very well that we cannot tamper with the revelation of original sin without undermining the mystery of Christ. (CCC 389)

It's like saying the church knows that if you mess with the concept of *original sin,* the whole Catholic edifice comes tumbling down.

Still, the transmission of original sin is a mystery that we cannot fully understand. But we do know by Revelation that Adam had received original holiness and justice not for himself alone, but for all human nature. By yielding to the tempter, Adam and Eve committed a personal sin, but this sin affected the human nature that they would then transmit in a fallen state. It is a sin which will be transmitted by propagation to all mankind, that is, by the transmission of a human nature deprived of original holiness and justice. And that is why original sin is called 'sin' only in an analogical sense: it is a sin 'contracted' and not 'committed'—a state and not an act. (CCC 404)

The Church's teaching on the transmission of original sin was articulated more precisely in the 5th century, especially under the impulse of St. Augustine's reflections against Pelagianism, and in the 16th century, in opposition to the Protestant Reformation. (CCC 406)

Still, the transmission of original sin is a mystery that we [the Catholic Church] cannot fully understand. (CCC 404)

Revelation gives us the certainty of faith that the whole of human history is marked by the original fault freely committed by our first parents. (Council of Trent 1513, CCC 390)

The real mystery is why theologians have not been able to undo the mystery they created in the 5th and16th centuries. They got in a bind they cannot remove themselves from.

But why did God not prevent the first man from sinning? . . . And St. Thomas Aquinas wrote, There is nothing to prevent human nature's being raised up to something greater, even after sin; **God permits evil in order to draw forth some greater good**. (CCC 412)

Here we have the official Catholic Catechism and St. Thomas Aquinas preaching . . . and basically saying, *the end justifies the means? But the end does not justify the means.* (CCC 1753) And don't forget about, *One may not do evil so that good may result from it.* (CCC 1756)

But *A sin is always connected with an individual person, who knowingly and willingly agrees to something evil.* (YOUCAT 320) (Did you knowingly and willingly agree to Adam and Eve's original sin? I sure didn't.) Bill Donohue of the Catholic League understands bishops cannot be held responsible for decisions made by preceding bishops. In a different setting: *But in this regard, bishops have no rival: sitting bishops are being held responsible to an absurd degree for the rulings made by those who preceded them.* (Bill Donohue, October 25, 2013)

But . . . God still holds all of humanity responsible for Adam and Eve's original sin?

Okay, if you participate in a crime, you are responsible. That's easy. The CCC agrees:

Sin is a personal act. Moreover, we have a responsibility for the sins committed by others when we cooperate in them: by participating directly and voluntarily in them. . . . (CCC 1868)

I didn't participate with Adam and Eve. I wasn't there.

A sin committed under compulsion, moreover, would be no sin, because sin always involves a free decision. (YOUCAT 69)

I made no free decision to commit the original sin.

By his sin Adam, as the first man, lost the original holiness and justice he had received from God, not only for himself but for all human beings. (CCC 416)

Actually, God defined/introduced sin and decided that all humanity would be affected by it.

God chose and blessed us for a purpose: to be holy and blameless in His sight. He chose us, each of us to be witnesses of His truth and His justice in this world. He created the world as a beautiful garden and asked us to care for it. But through sin, man has disfigured that natural beauty; through sin, man has also destroyed the unity and beauty of our human family, creating social structures which perpetuate poverty, ignorance and corruption. (Pope Francis, Vatican City, VIS, January 18, 2015)

Therefore sin is responsible for poverty and ignorance? Could this mean that since God introduced the original concept of sin, then God is responsible for poverty and ignorance?

Are we responsible for the sins of other people? No, we are not responsible for other people's sins. (YOUCAT 319, CCC 1868)

This contradicts a core Christian belief in Adam and Eve's sin. Is this an official evolution of Catholic theology?

Let's quickly review some of the mysteries and miracles needed to make the Adam and Eve story somewhat plausible.

1. Who created Adam and Eve?
2. Who decided knowledge of good and evil was bad?
3. Who taught Adam and Eve about obedience?
4. Who gave the serpent the ability to talk?
5. Who gave the snake the ability to talk in Adam and Eve's language?
6. Who should have taught Adam and Eve not to listen to talking snakes?

7. Who decided that disobediently eating an apple deserved a severe punishment?

8. Who decided all humanity should be affected/punished by the actions of Adam and Eve?

9. Who decided that I am affected by their actions? I wasn't there.

10. Who introduced death via a radical miraculous change in physiology after the fall?

11. Who performed all the miracles to make this story work?

12. Who knew all the above before creating Adam and Eve?

The obvious answer to all these questions is *God*. And it does look to Christians like God is pulling the strings and Adam, Eve, and all of humanity are along for the ride. God created the characters with full knowledge they would sin and God knew he would propagate that original sin. Doesn't this seem much the same as the miracles and mysteries that are introduced to make Santa Claus believable? And this is in Chapters 1 and 2 of Genesis, the very first book of the Biblical Old Testament!

Remember in *Alice in Wonderland* where the queen declares that she once believed six impossible things before breakfast? That's what's happening here. In fact, we are being asked to believe twice as many impossible things. We are expected to accept that knowledge of good and evil is bad. Do you really believe that? Is not Christianity all about teaching what is good and evil? Next, we are expected to believe eating an apple without permission is infinitely bad, for everybody. How could Adam and Eve know disobedience was bad? *Bad* did not exist till they ate the apple. And thirdly, we are led to believe the first two infractions are so serious that all of humanity must suffer eternal damnation unless they opt for the Jesus exemption.

If you can believe all these impossible things at the beginning, the rest of the Bible should be easy. Then again, the next story may present some serious challenges.

Noah's Ark and the Flood

This story, featured prominently in many a children's book, could actually be termed the greatest act of genocide ever, the mass murder of many innocents by God.

Let's take a modern perspective: How does the flood story read today?

According to Genesis 7:19–20, the mountain tops were submerged, with water covering them fifteen cubits (about 45 feet) high. We know that the world's highest mountain is actually about 29,000 feet above sea level. We also know that at that altitude the air is extremely cold. It might snow, but it never rains at 29,000 feet. At that altitude the air is also thin enough that breathing is almost impossible for humans, even for a short time. It is fair to assume that would be a problem for the ark's animals and plants, especially for the duration of a year. Further, what was the origin of all the water? And what happened to all of it after a year? Where did it drain to?

Although more than 99 percent of all species that ever lived on the planet are estimated to be extinct, there are currently 10–14 million species of life on the Earth. (Wikipedia)

Next, consider the enormous number of animals and plants that would have needed to be preserved on the ark. There were literally millions of species. They needed a wide range of living conditions (warm, cold, humid, and dry) and a broad variety of food to thrive, enough to last over a year. How long would it take your family to catch alive a male/female pair of the following half dozen creatures and transport them to the ark: bald eagle, monarch butterfly, bison, polar bear, okapi, and northern hairy-nosed wombat? (If you want a slightly more complete list, one that indicates the scope of such an assignment, go to Leviticus 11.) To accomplish what Noah is supposed to have done, you would need to know how to identify all the plants and animals on earth and their sexes at a distance and how to catch, transport, house and feed them.

How long do you think it would take to load a few million creatures into the ark even if they were lined up two-by-two holding hands? I've witnessed what it takes to load a few heifers and hogs onto a truck and it's not pretty nor easy.

Then there's the challenge of caring for them all once they are aboard the ark. If you have ever taken care of a pet or plant, you can imagine the knowledge, time, and effort that would be needed to care for millions of animals and plants, most of them strange and new to you. Almost every species needed a different setting (nest, pen, cage, etc.) with different living conditions. You can't optimize the space by placing the males and females together. Most needed different foods, including the kind that might be provided by chowing down on other ark occupants. (Lions and tigers need fresh meat, for instance. Do you keep extra live animals for them to eat or do you dry meat ahead of time and then reconstitute it?) Knowing what each species needed to survive had to be like knowing what every child in the world wants from Santa.

Then again, the Bible isn't very consistent on exactly what Noah collected. In one spot, it says he collected *two of every sort.* (Genesis 6:20) A few verses later, it says God commanded him to take *seven pairs of all clean animals* and *a pair of the animals that are not clean.* (Genesis 7:3) As someone who grew up on a dairy farm, I can tell you that attempting to create a herd starting with only two creatures isn't going to work well. Even seven pairs might not be enough. Check with the biologists about the need for genetic diversity. It's the same reason siblings and first cousins are counseled not to marry and have children.

Consider an old-fashioned, small, upper-Midwest farm, in its way a kind of mini-ark. If you've driven by one, you've probably noticed the barns, silos and granaries. These buildings hold the hay, straw, and grain to feed a few cows, calves, hogs and chickens for about six winter months. The manure created by these animals during that time makes a huge pile when shifted outside by the dint of much human labor. The water the animals consume is pumped from a well every day. It takes a family to take care of these animals. And this is just for a few domesticated farm animals.

Need another way to attempt to comprehend the task of taking care of a few million species for over a year on a boat? Consider the modern zoo. The cost in labor, food, storage, veterinary needs, safety, etc., for even a few thousand species is often more than a large city or state cares to sustain. You can check with your local zoo to verify the cost, specialists needed, and staff to care for their sample population.

The Noah story postulates a small family could build the ark, gather plants and animals from all over the earth, transport them to the ark, and amass the food needed to sustain everything for a year. All without modern tools and machines. Is this beginning to feel like flying reindeer and delivering toys all over the world in one night?

After a year, the ark—running out of food—settles onto the dry ground of a mountaintop (never mind where all the water went) and Noah is desperately ready to let the animals out. The problem is that at this point there isn't a living plant or animal anywhere on the whole earth. Remember, the whole point of the flood was to kill everything on Earth.

And, behold, I, even I, do bring a flood of waters upon the earth, to destroy all flesh, wherein is the breath of life, from under heaven; and every thing that is in the earth shall die. (Genesis 6:17-20)

For yet seven days, and I will cause it to rain upon the earth forty days and forty nights; and every living substance that I have made will I destroy from off the face of the earth. (Genesis 7:4)

Thus He blotted out every living thing that was upon the face of the land, from man to animals to creeping things and to birds of the sky, and they were blotted out from the earth; and only Noah was left, together with those that were with him in the ark. The water prevailed upon the earth one hundred and fifty days. (Genesis 7:23–24)

That seems pretty clear. But Noah is checking, just to be sure. There is nothing outside the ark to eat and there won't be for months or even years. The first time he sends out a dove, it returns because the waters are *still on the face of the whole earth*. He sends it out again seven days later and . . .

And the dove came in to him in the evening; and, lo, in her mouth was *an olive leaf pluckt off: so Noah knew that the waters were abated from off the earth.* (Genesis 8:11)

Today if standing water covers a dip in the ground for only a few months and then dries, the ground is bare of plants. But here an olive tree not only somehow was seeded (remember all the extant plant material is on the ark), but it managed to grow enough to leaf out in just a week. Talk about Miracle Grow!

Recall, too, that even if there is an olive tree somewhere, many of the animals on the ark are carnivorous. Imagine the chaos and carnage that could ensue when these hungry animals are released to a bare Earth void of any other creatures.

These hungry animals must have cried, too, when they saw Noah stop to sacrifice one each of the not-clean animals he fed for a year. Why did he preserve them in the first place? To enable this sacrifice? What is the sacrifice for? God set up the whole flood thing, it worked, and now he needs some kind of tribute? Further, if one of each pair of not-clean animals—camels, rabbits, pigs, moles, mice, lizards, (Leviticus 11:29–31) as well as four-footed animals with paws such as cats, dogs, bears, lions, and tigers—is now sacrificed, how will the other animal of the pair multiply now? Today we figure if a species gets down to two creatures, it's on the way to extinction.

And who's going to plant all the plants from the ark? How long will it take for the seeds in a small oak cone to become a small oak forest? How long will it take for a pair of American eagles to become a flock of American eagles, especially when there is nothing to eat?

And God blessed Noah and his sons, and said unto them, be fruitful, and multiply [author's note: again, incest seems the only option here], and replenish the earth. (Genesis 9:1, Genesis 28:3, Genesis 35:11, and Genesis 48:4).

He promises that he will never again *curse the ground because of man* or *destroy every living creature as I have done.* (Genesis 8:21)

And I will establish my covenant with you, neither shall all flesh be cut off any more by the waters of a flood; neither shall there any more be a flood to destroy the earth. And God said, This is the token of the covenant which I make between me and you and every living creature that is with you, for perpetual generations. I set my bow in the cloud, and it shall be a sign of the covenant between me and the earth. (Genesis 9:11–13)

God made an everlasting covenant with Noah and with all living beings. [See Genesis 9:15] *It will remain in force as long as the world lasts.* (CCC 71)

What a noble thing to do. The rainbow symbol is pretty, too. In the process of making his promise, though, God also mentions that *the imagination of man's heart is evil from his youth.* (Genesis 8:21) And he fails to mention that his laying waste to the earth is not going to get humanity off the hook for Adam and Eve's transgressions. Or that he knows that he's going to have to sacrifice his son at some point in order to save those few who enter by *the narrow gate.* (See Matthew 7:13–14)

The Catholic Church regularly compares itself to Noah's ark. Because only Noah's ark saves people.

According to another image dear to the Church Fathers, she [The Catholic Church] is prefigured by Noah's ark, which alone saves from the flood. (CCC 845)

How are we to understand this affirmation, often repeated by the Church Fathers? Re-formulated positively, it means that all salvation comes from Christ the Head through the Church which is his Body . . . (CCC 846)

This is not the only ark story, by the way—there are a number that are even older than this one. Sumerians had Ziusudra instead of Noah. In Akkad, there was Atrakhasis; in Babylon, Uta-Napisthim; in Greece, Deucalion; in Armenia, Xisuthros; in India, Menu; in Egypt, Osiris; and in Mexico, Nata.

Noah and the flood are central, too, to some of the most interesting claims in creationism, particularly in its arguments against evolution. According to creationism, all fossils should lie in one layer of sediment—now stone—laid down during the Flood. (See Genesis 7:21–23) Evolution argues for different fossils being found in many different layers of sediment/stone (deeper = older).

Creationists argue that the Flood would first engulf marine animals, then slow lowland creatures such as reptiles, while wily and speedy humans escaped to the hilltops. To a creationist, this adequately explains the order in which fossils occur in the geologic column. A scientist might test the mobility hypothesis by examining how well it explains the fact that flowering plants don't occur in the fossil record until early in the Cretaceous era. A scenario with magnolias (a primitive plant) heading for the hills, only to be overwhelmed along with early mammals, is unconvincing. And when marine fossils are found in many places above those of land animals and plants, the victim habitat apologetic loses all credibility, too. (See Robert J. Schadewald, *Six 'Flood' Arguments Creationists Can't Answer*)

Evolution theory predicts one will never find, say, a rabbit fossil in the rock layer and time epoch called the Proterozoic, the geological period just before the proliferation of complex life. And thus far nobody has.

Remember this bit from Part III? *God's action is the model for human action.* (CCC 2172) God killed the people he created because he didn't like how they *turned out*. I'm pretty sure parents can't do that today. And didn't do it that often in the past.

Okay, so why did God do it?

And God said unto Noah, The end of all flesh is come before me; for the earth is filled with violence through them; and, behold, I will destroy them with the earth. (Genesis 6:13)

There was too much violence, so God decides to kill everything violently. Well, almost everything.

How could an omniscient God not know his creation was destined to be a disaster and a disappointment? And if he did know a deadly flood lay in its future, doesn't it seem evil to continue what he was doing? Arguably, one can claim God is more merciless than Satan.

In the flood, men, women, and innocent children perished with no explanation, no chance at obtaining forgiveness. The devil begs, pleads, and cajoles to entice people to follow his evil suggestions. God just—*Poof!*— destroys everything. This phenomenon repeats often in the Bible.

The New Testament

The Birth of Christ

The earliest Gospel writings date to about 50 years after Jesus's birth, which is placed between 4 BCE and 6 CE by Bible scholars. No contemporary historical records exist.

The Jesus story begins here:

And in the sixth month the angel Gabriel was sent from God unto a city of Galilee, named Nazareth, to a virgin espoused to a man whose name was Joseph, of the house of David; and the virgin's name was Mary. And the angel came in unto her, and said, Hail, thou that art highly favoured, the Lord is with thee: blessed art thou among women. And when she saw him, she was troubled at his saying, and cast in her mind what manner of salutation this should be. And the angel said unto her, Fear not, Mary: for thou hast found favour with God. And, behold, thou shalt conceive in thy womb, and bring forth a son, and shalt call his name Jesus. (Luke 1:26-31)

Divinely inspired pregnancies were common in Greek and Roman mythology, so it was easy for the early Christians to accept and pass along the divine Jesus conception. It was so common it even shows up in Genesis 6:1–4, where multiple *sons of God* impregnate the *daughters of men.*

The Vatican explains that God chose to be born *in a human family, that He Himself had formed.* (Pope Francis, December 17, 2014) This is billed as emphasizing the importance of the good family, but what a model family! Gabriel not only does not first ask Mary if she wants to get pregnant (today we would call such an event a rape), but he also neglects to clue Joseph in on what's going on. However you look at it, God put Mary in a pretty difficult spot. Two thousand years ago (and maybe even today) in that part of the world, getting pregnant without her husband's involvement would require that she be stoned to death.

Joseph was, of course, bothered by what Mary told him. He was considering *putting her away* (divorcing her) until the mystery of non-fatherhood was revealed to him belatedly via another miracle:

But while he thought on these things, behold, the angel of the Lord appeared unto him in a dream, saying, Joseph, thou son of David, fear not to take unto thee Mary thy wife: for that which is conceived in her is of the Holy Ghost. (Matthew 1:20)

Not to worry, Joseph. The Holy Ghost did it. And dreams are true. If you believe your dreams, you can believe anything. In any case, you can regard Jesus as having been conceived *of the seed of David according to the flesh.* (Romans 1:3)

Why does the Bible bother including paternal genealogic information on Jesus since clearly Joseph was not the father of Jesus? Probably because of the promises God made to David that he would see his descendant sitting on his throne, the throne of David in Jerusalem, and that his Kingdom would continue forever. (2 Samuel 7:13–16, 2 Samuel 7:26)

The Gospel writers no doubt intended to signal that Christ was that promised one, the *king of the Jews*, but the *evidence* they offer up is muddled and contradictory. Matthew 1:16 names Jacob as the paternal grandfather of Jesus; Luke 3:23 says it was Heli. An even more dramatic set of Biblical contradictions occurs between Matthew 1:1–17, which traces the genealogy from Abraham to Joseph to Jesus, and Luke 3:23–38, which offers a genealogy from God to Adam to Joseph to Jesus. Does God have DNA to pass along? Did Jesus? A few more miracles are needed here to make this work.

The Incarnation is therefore the mystery of the wonderful union of the divine and human natures in the one person of the Word. (CCC 483)

A number of other mysteries surround the birth and early life of Jesus.

Exactly where Jesus was born is still an open question among scholars. The Bible's a little fuzzy about it too.

Matthew 2:11 says it was in a house.

Luke 2:7 says it was in a manger.

When was Jesus born?

. . . when Cyrenius was governor of Syria. (6-12 CE, Luke 2:2)

. . . in the days of Herod the king. (37-4 BCE, Matthew 2:1)

Why was Jesus born in Bethlehem, rather than Galilee?

And it came to pass in those days, that there went out a decree from Caesar Augustus, that all the world should be taxed. . . . And all went to be taxed, every one into his own city. And Joseph also went up from Galilee, out of the city of Nazareth, into Judaea, unto the city of David, which is called Bethlehem, because he was of the house and lineage of David. (Luke 2:1-4)

While there are many Roman records from this era, there is no record of a Roman, much less an Empire-wide, tax/census. This is much like Santa visiting every house in one night. The Roman Empire was too big to have everyone go home (to which ancestor?) to be counted. The time and cost would have been prohibitive.

In Matthew it is written that the baby Jesus was threatened by Herod, but Herod was dead by 4 BCE. And, while there are contemporary written records of Herod, there are none of Jesus. Everything we know of Jesus has been provided by Christians. And Christians have been so split and divided regarding the details of the life of Jesus that various councils had to come together over the course of ten centuries to come up with something that everyone could (almost) agree on.

It's also worth noting that there are problems with Christmas, the annual celebration of the birth of Jesus.

Before about 350 CE, the Winter Solstice was an established holiday, an occasion in Rome during which people celebrated the birth of their gods with decorated trees, feasts, gift-giving and family get-togethers. Sound familiar? In *Jesus of Nazareth,* Pope Benedict XVI claims 336

CE as the first year that the church calendar showed December 25 as the year of Jesus's birth. (To help sway the Romans to Christianity, Theodosius ordered all pagan temples should be destroyed around this time.) After 380 CE the Christians made the December 25th solstice more or less their holiday exclusively.

There are problems with the story of the Christmas star as well. First of all, kings from three different locations, each following the same star, would end up in parallel paths, instead of converging as the story says. Also, due to the earth's rotation, stars move all night, every night, from the eastern horizon to the western horizon. In order to make it possible to *follow* a single, albeit bright, one, a few more miracles would have to happen.

Finally, a word about the so-called *war on Christmas*. It could be argued that it's something that dates back to colonial days: Between 1659 and 1680 the Puritans in the Massachusetts Bay Colony forbade the celebration of Christmas. Apparently Christmas was already a problem before the United States was formed. Alabama was the first state to recognize Christmas as a legal holiday—in 1836. It didn't become a federal holiday until 1894.

The Death of Christ—the Crucifixion

Was the death of Christ on the cross essentially a case of *suicide by cop?* That is, a method in which an individual deliberately behaves so as to provoke a lethal response from a public safety officer, law enforcement officer, or other legitimately armed individual?

Did Jesus know that he would die when He entered Jerusalem? Yes. Three times Jesus had predicted his suffering and death before consciously and voluntarily going to the place of his passion and his resurrection. (YOUCAT 94, Luke 9:51)

Jesus went up to Jerusalem voluntarily, knowing well that there he would die a violent death because of the opposition of sinners. (CCC 569)

Plus, he was an omniscient God, wasn't he?

No man taketh it from me, but I lay it down of myself. I have power to lay it down, and I have power to take it again. This commandment have I received of my Father. (John 10:18)

And Jesus going up to Jerusalem took the twelve disciples apart in the way, and said unto them, Behold, we go up to Jerusalem; and the Son of man shall be betrayed unto the chief priests and unto the scribes, and they shall condemn him to death. (Matthew 20:17-18)

Actually, if this were a true direct quote, wouldn't Jesus have said, *And they shall condemn me to death?*

The violent death of Jesus did not come about through tragic external circumstances. Jesus was 'delivered by the determinate counsel and foreknowledge of God.' (Acts 2:23, YOUCAT 98)

What is the punishment nowadays for premeditating and then effecting the killing of someone? Life in prison or the death penalty. God had a plan. He knew what he was doing:

Him, being delivered by the determinate counsel and foreknowledge of God, ye have taken, and by wicked hands have crucified and slain. . . . (Acts 2:23)

And he is the propitiation for our sins: and not for ours only, but also for the sins of the whole world. (1 John 2:2)

The rejection and murder of God's only Son, caused by the sins of all men. . . . (CCC 312)

Jesus freely offered himself for our salvation. Beforehand, during the Last Supper, he both symbolized this offering and made it really present: 'This is my body which is given for you.' (CCC 621)

For God so loved the world, that He gave His only begotten Son, that whosoever believeth in him should not perish, but have everlasting life. (John 3:16)

This willingness to sacrifice a son was not unique to Christian or Catholic Biblical beliefs, of course. Remember Abraham preparing to sacrifice Isaac? This story fits a common worldwide history of human sacrifices, which was practiced by Phoenicians and Carthaginians, Celtic Druids, Goths, Manchus, ancient Hawaiians, Mayans, Aztecs,

Incas, and early American Natchez, Pawnee, and Iroquois societies. Thus it was easily accepted that Jesus could be sacrificed to God so that God would be placated and have no need to punish the humans he created for violating the rules he created. There's also a certain Munchausen Syndrome by Proxy aspect to the crucifixion. After all, isn't God essentially harming (actually sacrificing, killing) his son so that more humans can get into heaven to adore God?

When the hour had come for him to fulfill the Father's plan of love, Jesus allows a glimpse of the boundless depth of his filial prayer, not only before he freely delivered himself up . . . (CCC 2605)

When Christ died on the cross, God went P*oof! Everybody is reconciled.* Couldn't God just have skipped the cross thing and gone *Poof! Everybody is reconciled*? *After all, There is no sin that God cannot forgive! None!* (Pope Francis, VIS March 12, 2015) Also, if it's true that God can forgive any sin, then why the need for hell? And if God hears prayers, then why not address your confessions to God directly, rather than through a priest?

This sacrifice of Christ is unique; it completes and surpasses all other sacrifices. First, it is a gift from God the Father himself, for the Father handed his Son over to sinners in order to reconcile us with himself. At the same time it is the offering of the Son of God made man, who in freedom and love offered his life to his Father through the Holy Spirit in reparation for our disobedience. (CCC 614)

This quote from the Catholic Catechism claims, *the Father handed his Son over to sinners in order to reconcile us with himself.* Let's look at that a little more closely. Try this analogy: I'm God. Your great grandparents disobeyed me and ate an apple, so I decided that all of humanity must suffer eternal damnation. Now, though, I'm going to give you my son so that you can kill him. That will make up for everything—you and all your relatives before you and after you will be reconciled with me for eating that apple. Does that work for you?

This type of thinking may have been valid 2,000 years ago, but now? Dying and coming back to life a few days later may have been a gift to humanity, but was it really a sacrifice?

Why did Jesus have to redeem us on the Cross, of all places? . . . Christ our Redeemer, chose the Cross so as to bear the guilt of the world and to suffer the pain of the world. . . . (YOUCAT 101)

The catechism doesn't really answer its own question. A weird attempt at clarity.

But if we walk in the light, as he is in the light, we have fellowship one with another, and the blood of Jesus Christ his Son cleanseth us from all sin. (1 John 1:7)

Blood cleanses us from sin?

The Second Vatican Council clearly states that the Jews as a people cannot be charged with any collective guilt for the crucifixion of Christ. (YOUCAT 135)

That's good to know. Are there any other things we should try to straighten out while we're at it? How about some of these contradictions?

How many times did the cock crow when the priests were interrogating Jesus?

Then began he to curse and to swear, saying, I know not the man. And immediately the cock crew. And Peter remembered the word of Jesus, which said unto him, Before the cock crow, thou shalt deny me thrice. And he went out, and wept bitterly. (Matthew 26:74-75)

And the second time the cock crew. And Peter called to mind the word that Jesus said unto him, Before the cock crow twice, thou shalt deny me thrice. And when he thought thereon, he wept.(Mark 14:72)

And Peter said, Man, I know not what thou sayest. And immediately, while he yet spake, the cock crew. And the Lord turned, and looked upon Peter. And Peter remembered the word of the Lord, how he had said unto him, Before the cock crow, thou shalt deny me thrice. (Luke 22:60-61)

Jesus answered him, Wilt thou lay down thy life for my sake? Verily, verily, I say unto thee, The cock shall not crow, still thou hast denied me thrice. (John13:38)

Peter then denied again: and immediately the cock crew. (John 18:27)

Who carried the cross?

Jesus alone (John 19:17)

Jesus with help from Simon of Cyrene (Mark 15:21, Luke 23:26, Matthew 27:32)

What was the inscription on the cross?

Jesus of Nazareth, King of the Jews (John 19:19)

The King of the Jews (Mark 15:26)

On what day was he crucified?

The day before the Passover (John 19:14–16)

The day of the Passover (Mark 14:12)

At which hour was Jesus crucified?

The third hour (Mark 15:25)

The sixth hour (John 19:14)

Peter and Paul state, however, that Jesus was hanged: *The God of our fathers raised up Jesus, whom ye slew and hanged on a tree.* (Acts 5:30)

What did they give Jesus to drink?

Vinegar (John 19:29-30)

Wine with myrrh (Mark 15:23)

What were Jesus's last words?

And about the ninth hour Jesus cried with a loud voice, saying, 'Eli, eli, lama sabachthani?' that is to say, 'My God, my God, why hast thou forsaken me?' . . . Jesus, when he cried again with a loud voice, yielded up the ghost. (Matthew 27:46-50)

And when Jesus had cried with a loud voice, he said, 'Father, unto thy hands I commend my spirit': and having said thus, he gave up the ghost. (Luke 23:46)

When Jesus therefore had received the vinegar, he said, It is finished: and he bowed his head, and gave up the ghost. (John 19:30)

Was the temple curtain ripped before or after Jesus died?

Jesus, when he had cried again with a loud voice, yielded up the ghost. And, behold, the veil of the temple was rent in twain from the top to the bottom; and the earth did quake, and the rocks rent. (Matthew 27:50-51)

And the sun was darkened, and the veil of the temple was rent in the midst. And when Jesus had cried with a loud voice, he said, Father, into thy hands I commend my spirit: and having said thus, he gave up the ghost. (Luke 23:45-46)

And the sun was darkened, and the veil of the temple was rent in the midst. And when Jesus had cried with a loud voice, he said, Father, into thy hands I commend my spirit: and having said thus, he gave up the ghost. (Mark 15:37-38)

Looks like someone copied from somebody else here. If I were an English teacher grading tests who found these last two quotes, I'd definitely be suspicious. The fact that the wording is identical in two *different* accounts does not strengthen their claim.

Whose sepulcher was it anyway?

And were carried over into Sychem, and laid in the sepulchre that Abraham bought for a sum of money of the sons of Emmor the father of Sychem. (Acts 7:16)

When the event was come, there came a rich man of Arimathaea, named Joseph, who also himself was Jesus' disciple . . . and begged the body of Jesus . . . And laid it in his own new tomb. (Matthew 27:57-60)

Note: Today Calvary, Golgotha, Talpiot, Kashmir, and Shingo, Japan, all claim to be the site of Jesus's tomb.

Who anointed Christ's body?

A group of women came to anoint after the burial. (Mark 16:1-3)

Nicodemus and Joseph of Arimathaea anointed before the burial. (See John19:38-40)

In case you're wondering—scientific analysis has shown that the Shroud of Turin is a 14th century forgery. It's easy today, without touching it, to prove it's a forgery. But, as of June 21, 2015, the Vatican still supports that it has value.

The Shroud is a cloth, but it is above all an image. . . . This image tells us of Jesus, in an immediate way, before science can offer its version and before faith reveals that it is Jesus. All this does not mean that the Shroud is certainly the sheet brought by Joseph of Arimathea below the cross, but certainly anyone who looks at it will find that it immediately recalls this story. (Roberto Gottardo, president of the diocesan Commission for the Shroud, VIS June 21, 2015)

How long was Jesus in the tomb?

The answer depends on where you look. Matthew 12:40 has Jesus prophesying that he will spend *three days and three nights in the heart of the earth,* and Mark 10:34 has *And they shall mock him, and shall scourge him, and shall spit upon him, and shall kill him: and the third day he shall rise again.*

But Jesus was crucified Friday afternoon and risen by Sunday morning, a period of not more than two full days.

How many women visited Christ's tomb?

One, Mary Magdalene, visits the tomb. (John 20:1)

Two women, Mary Magdalene and another Mary visit the tomb. (Matthew 28:1)

Three women, Mary Magdalene and another Mary and Salome visit the tomb. (Mark 16:1)

Five or more women visit the tomb. (Luke 24:10)

Who was at the tomb when they arrived?

One angel—sitting on the stone (Matthew 28:2-7)

One young man—sitting inside on the right (Mark 16:5)

Two men—standing inside (Luke 24:4)

Two angels—sitting on each end of the bed (John 20:12)

When the women arrived, was the tomb open or closed?

And they found the stone rolled away from the sepulchre. (Luke 24:2, Mark 16:4, John 20:1)

And, behold, there was a great earthquake: for the angel of the Lord descended from heaven, and came and rolled back the stone from the door, and sat upon it. (Matthew 28:2)

When did Mary first see Jesus?

Before she returned to the disciples (Matthew 28:9, Mark 16:9-10)

After she returned to the disciples (John 20:2,14)

Could Jesus be touched after the resurrection?

Yes (Matthew 28:9, Luke 24:39)

No (John 20:17)

After the resurrection:

Jesus appeared, to prove his resurrection. (1 Corinthians 15:3-8)

Jesus was already virtually among us. (Matthew 18:19-20)

And saintly zombies appeared?

And the graves were opened; and many bodies of the saints which slept arose, And came out of the graves after his resurrection, and went into the holy city, and appeared unto many. (Matthew 27:52-53)

One would think there would be a non-Biblical record of this astonishing event. There is not.

What happened at Jesus's first appearance?

Disciples worshipped, some doubted and were told to *Go preach.* (Matthew 28:17-20)

Disciples were reprimanded, told to *Go preach.* (Mark 16:14-19)

Jesus was absent, then *magically appeared,* disciples were reprimanded, they ate supper. (Luke 24:13-51)

Jesus passed through a solid door, disciples were happy and blessed by Jesus. (John 21:19-23)

Can God be seen?

Yes—*And the Lord spake to Moses face to face, as a man speaketh to his friend. . . .* (Exodus 33:11)

. . . for I have seen God face to face, and my life is preserved. (Genesis 32:30)

No—No man hath seen God at any time. . . . (John 1:18)

Thou canst not see my face; for there shall no man see me and live. (Exodus 33:20)

. . . whom no man hath seen nor can see . . . (1 Timothy 6:16)

Did Jesus stay more than a day on Earth after the resurrection?

No (Mark 16:19, Luke 24:50-52)

Yes—at least 8 days (John 20:26, 21:1-22)

Yes—at least 40 days (Acts 1:3)

Judas died how?

And he cast down the pieces of silver into the temple and departed, and went out and hanged himself. (Matthew 27:5)

And falling headlong, he burst asunder in the midst, and all of his bowels gushed out. (Acts 1:18)

Where did the ascension take place?

No ascension mentioned (Matthew, John, Paul)

In or near Jerusalem after supper (Mark 16:19)

In Bethany (close to Jerusalem) after supper (Luke 24:50-51)

He ascended from Mount of Olives (Acts 1:9-12)

Matthew, Mark, Luke, and Paul clearly contradict each other; it's obvious they can't all be accurate and true. It's possible that one of them may be correct. That means, though, that the other *inspired* writers must be wrong. Which is not very inspiring—unless one accepts the notion of divinely inspired contradictions.

Does the Bible mention any other ascensions into heaven besides that of Jesus?

And no man hath ascended up to heaven, but he that came down from heaven, even the Son of man which is in heaven. (John 3:13)

And it came to pass, as they still went on, and talked, that, behold, there appeared a chariot of fire, and horses of fire, and parted them both asunder; and Elijah went up by a whirlwind into heaven. (2 Kings 2:11)

Did those with Paul on the road to Damascus hear a voice identifying itself as that of Jesus at the moment of his conversion?

And the men which journeyed with him stood speechless, hearing a voice, but seeing no man. (Acts 9:7)

And they that were with me saw indeed the light, and were afraid; but they heard not the voice of him that spake to me. (Acts 22:9)

What kind of evidence exists about any of these events?

There are no Roman records of Pontius Pilate ordering the execution of Jesus. There are no Roman records of Jesus at all. Historians have offered no evidence that a god has ever raised anyone from the dead. Nevertheless . . .

Are there proofs for the Resurrection of Jesus? There are no proofs for the Resurrection of Jesus in the scientific sense. . . . (YOUCAT 106)

But, it *is* believable in the faith sense?

The mystery of Christ's resurrection is a real event, with manifestations that were historically verified, as the New Testament bears witness. (CCC 639)

So, everything in this book is true, because the writing in this book says so? And *mysteries* are real events. Believe it or else!

And if Christ has not risen, then our preaching is useless, and your faith is also useless. (1 Corinthians 15:14)

That if thou shalt confess with thy mouth the Lord Jesus, and shalt believe in thine heart that God hath raised him from the dead, thou shalt be saved. (Romans 10:9)

By his death and Resurrection, Jesus Christ has 'opened' heaven to us. (CCC 1026)

By the way, this is the final revelation, according to the Catholic Church:

God has revealed himself fully by sending his own Son, in whom he has established his covenant for ever. The Son is his Father's definitive Word; so there will be no further Revelation after him. (CCC 73)

The above will be a shock, I expect, to those popes and others who have with some regularity announced new Catholic dogma, i.e., articles of faith revealed by God which the church presents as necessary to be believed if one freely chooses to be a Catholic. For instance, the 20th century witnessed the introduction of the dogma of Assumption of Mary by Pope Pius XII in 1950. However, these beliefs were already held in some form or another within the church before their elevation to the dogmatic level. A movement to recognize Mary as a Mediatrix (mediator in the redemption offered by her son Jesus) and Co-Redemptrix (because she gave free consent to give life to the Redeemer, to share his life, to suffer with him under the cross, and to offer his sacrifice to God the Father for the sake of the redemption of mankind) is underway.

The End—The Second Coming

In the last book of the Bible, the Book of Revelation, we are told of the apocalypse whereby God, at the end of the world, finally sorts the good from the bad. The actual date of this doomsday has been predicted in recent times by many religious preachers such as Harold Camping (May & October 21, 2011), William Miller (May 21, 2011 & October 21, 2011), Fatima —3rd secret (2000 CE), Dorothy Martin (December 21, 1954), Jeane Dixon (February 4, 1962), Pat Robertson (Fall 1982). What they don't deem to understand is that God Himself promised, in writing, that it would happen nearly 2,000 ago. People alive then would live to see it.

Verily I say unto you, There be some standing here, which shall not taste of death, till they see the Son of man coming in his kingdom. (Matthew 16:28)

And he said unto them, Verily I say unto you, That there be some of them that stand here, which shall not taste of death, till they have seen the kingdom of God come with power. (Mark 9:1)

But I tell you of a truth, there be some standing here, which shall not taste of death, till they see the kingdom of God. (Luke 9:27)

For this we say unto you by the word of the Lord, that we which are alive and remain unto the coming of the Lord shall not prevent them which are asleep. For the Lord himself shall descend from heaven with a shout, with the voice of the archangel, and with the trump of God: and the dead in Christ shall rise first: Then we which are alive and remain shall be caught up together with them in the clouds, to meet the Lord in the air: and so shall we ever be with the Lord. (1 Thessalonians 4:15-17)

And it still hasn't happened.

More Conundrums

In addition to the mysteries surrounding the *mysteries* of the birth and death of Jesus, the Old and New Testament contains myriad other confusions. Some involve contradictions within different passages, as noted above. Others put forth assertions at odds with science, like the ones below, which seems to suggest that God was no mathematician.

Value of Pi

And he made a molten sea, ten cubits from the one brim to the other: it was round all about, and his height was five cubits: and a line of thirty cubits did compass it round about. (1 Kings 7:23)

This is interesting because mathematicians have known for centuries that it is impossible in the same circle for the circumference (thirty) and the diameter (ten) to simultaneously both be counting numbers. The ratio of circumference to diameter in 1 Kings 7:23 would be the number we all know and love as *three.* In reality the ratio of circumference to diameter of any circle is called pi and has a well-established value of approximately 3.14159 . . . an infinite non-repeating decimal. Perhaps God just preferred to round his answers to the nearest whole number?

Like father like son? Jesus doesn't seem to be much of a mathematician either, judging from his pronouncements on whether he is equal to or lesser than God.

I and my Father are one. (John 10:30)

Ye have heard how I said unto you, I go away, and come again unto you. If ye loved me, ye would rejoice, because I said, I go unto the Father: for my Father is greater than I. (John 14:28)

This helps explain why we can't take the Bible literally. The contradictions found in its precepts for how people should treat one another also explain why we shouldn't make *God's action the model for human action.* (See CCC 2172)

Violence is contrary to God's nature. (Pope Benedict XVI, September 13, 2006) Really? A video game based on Biblical stories and characters would likely be banned due to extreme violence. As you'll see below, the Old Testament contains the bulk of it, but the New Testament had its own elements of violence and division, too. In the Sermon on the Mount, Jesus taught that *Blessed are the peacemakers: for they will be called children of God,* (Matthew 5:9) but five chapters later in Matthew, you'll find this: *Think not that I am come to send peace on earth: I came not to send peace, but a sword. For I am come to set a man at variance against his father, and the daughter against her mother, and the daughter in law against her mother in law.* (Matthew 10:34-35)

You shall not kill. (CCC 2052)

The Biblical god personally kills a total of 371,186 people, not counting his slaughter of every living thing in Genesis 7. The Biblical god also orders the killing of a total of 1,862,265. (Barbara G. Walker, FFRF Freethought Today, March 2012)

And Judah went up; and the Lord delivered the Canaanites and the Perizzites into their hand: and they slew of them in Bezek ten thousand men. (Judges 1:4)

And it came to pass that night, that the angel of the Lord went out, and smote in the camp of the Assyrians an hundred fourscore and five thousand: and when they arose early in the morning, behold, they were all dead corpses. (2 Kings 19:35)

But, Pope Francis claims the Gospel values are *love, justice, peace and sobriety.* (VIS 03-12-2015—Year XXII—Num. 217)

Don't forget the raping and the pillaging. That could be pretty darn profitable.

And they utterly destroyed all that was in the city, both man and woman, young and old, and ox, and sheep, and ass, with the edge of the sword. But Joshua had said unto the two men that had spied out the country, Go into the harlot's house, and bring out thence the woman, and all that she hath, as ye sware unto her. And the young men that were spies went in, and brought out Rahab, and her father, and her mother, and her brethren, and all that she had; and they brought out all her kindred, and left them without the camp of Israel. And they burnt the city with fire, and all that was therein: only the silver, and the gold, and the vessels of brass and of iron, they put into the treasury of the house of the Lord. (Joshua 6:21-24)

And the booty, being the rest of the prey which the men of war had caught, was six hundred thousand and seventy thousand and five thousand sheep, And threescore and twelve thousand beeves, And threescore and one thousand asses, And thirty and two thousand persons in all, of women that had not known man by lying with him. And the half, which was the portion of them that went out to war, was in number three hundred thousand and seven and thirty thousand and five hundred sheep: And the Lord's tribute of the sheep was six hundred and threescore and fifteen. And the beeves were thirty and six thousand; of which the Lord's tribute was threescore and twelve. And the asses were thirty thousand and five hundred; of which the Lord's tribute was threescore and one. And the persons were sixteen thousand; of which the Lord's tribute was thirty and two persons. (Numbers 31:32-40)

What kind of relationship should parents and children have?

For whomever the Lord loves, he chastises. And every son whom he accepts, he scourges. Persevere in discipline. God presents you to himself as sons. But what son is there, whom his father does not correct? (Hebrews 12:6–7)

He who spares the rod hates his son. But he who loves him urgently instructs him. (Proverbs 13:24)

Do not be willing to take away discipline from a child. For if you strike him with the rod, he will not die. (Proverbs 23:13)

The rod and its correction distribute wisdom. But the child who is left to his own will, brings shame to his mother. (Proverbs 29:15)

What about women? How should they be treated? The Biblical record isn't very pretty.

Lot offers two daughters to the mob outside to save two angels disguised as men. (See Genesis 19:8) God advises Moses to keep the virgins as spoils of war and kill everyone else. (See Numbers 31:18) To save a male guest in his house a man sacrifices his concubine to the mob outside; after they *abuse* her all night, he cuts her body into twelve pieces and scatters them around Israel. (Judges 19:25) God gives David's wives to his neighbor. (2 Samuel 12:11)

On the other hand, women make pretty good sex objects.

A bundle of myrrh is my well-beloved unto me; he shall lie all night betwixt my breasts. (Song of Solomon 1:13)

Thy two breasts are like two young roes that are twins. (Song of Solomon 7:3)

I am a wall, and my breasts like towers: then was I in his eyes as one that found favour. (Song of Solomon 8:10)

Let her be as the loving hind and pleasant roe; let her breasts satisfy thee at all times; and be thou ravished always with her love. (Proverbs 5:19)

What about wives? They get treated better, don't they?

To the woman, he also said: I will multiply your labors and your conceptions. In pain shall you give birth to sons, and you shall be under your husband's power, and he shall have dominion over you. (Genesis 3:16)

Wives, be submissive to your husbands, as is proper in the Lord. (Colossians 3:18)

Wives should be submissive to their husbands, as to the Lord. For the husband is the head of the wife, just as Christ is the head of the Church. He is the Savior of his body. Therefore, just as the Church is subject to Christ, so also should wives be subject to their husbands in all things. (Ephesians 5:22-24)

Let a woman learn in silence with all subjection. For I do not permit a woman to teach, nor to be in authority over a man, but to be in silence. For Adam was formed first, then Eve. And Adam was not seduced, but the woman, having been seduced, was in transgression. (1 Timothy 2:11-14)

So I want you to know that the head of every man is Christ. But the head of woman is man. Yet truly, the head of Christ is God. (1 Corinthians 11:3)

For man is not of woman, but woman is of man.

And indeed, man was not created for woman, but woman was created for man. (1 Corinthians 11:8-9)

Okay, what about homosexuals? Should you banish them?

And Asa did that which was right in the eyes of the Lord, as did David his father. And he took away the sodomites out of the land, and removed all the idols that his fathers had made. (1 Kings 15:11-12)

Or should you kill them?

If a man lies with a male as he lies with a woman, both of them have committed an abomination. They shall surely be put to death. Their blood shall be upon them. (Leviticus 20:13)

Same goes for those who engage in bestiality. And for their victims.

Whoever will have committed sexual acts with any animal or cattle, he shall die a death. Likewise, you shall slay the beast. (Leviticus 20:15)

What about incest?

And it came to pass on the morrow, that the firstborn said unto the younger, Behold, I lay yesternight with my father: let us make him drink wine this night also; and go thou in, and lie with him, that we may preserve seed of our father. And they made their father drink wine that night also: and the younger arose, and lay with him; and he perceived not when she lay down, nor when she arose. Thus were both the daughters of Lot with child by their father. (Genesis 19:34–36)

How should all other people be treated?

And thou shalt love the Lord thy God with all thy heart, and with all thy soul, and with all thy mind, and with all thy strength: this is the first commandment. And the second is like, namely this, Thou shalt love thy neighbor as thyself. There is none other commandment greater than these. (Mark 12:30–31)

Ye have heard that it hath been said, An eye for an eye, and a tooth for a tooth: But I say unto you, That ye resist not evil: but whosoever shall smite thee on thy right cheek, turn to him the other also. (Matthew 5:38–39)

But I say unto you which hear, Love your enemies, do good to them which hate you, Bless them that curse you, and pray for them which despitefully use you. And unto him that smiteth thee on the one cheek offer also the other; and him that taketh away thy cloak forbid not to take thy coat also. (Luke 6:27–29)

As always, there are exceptions to these rules.

For I say unto you, That unto every one which hath shall be given; and from him that hath not, even that he hath shall be taken away from him. But those mine enemies, which would not that I should reign over them, bring hither, and slay them before me. (Luke 19:26-27)

Pope Francis seems, surprisingly, more comfortable than you might expect with this latter alternative:

In a quite astonishing response to the Charlie Hebdo massacre in Paris, the Pope not only defended those who use violence to deny others the right to criticise religion but reversed centuries of Christian official teaching, and said that people who insult religion should expect to be punched, just as someone should expect to be punched for insulting someone's mother. (Rosa Rubicondior, January 15, 2015, based on BBC News, January 15, 2015)

That's right. Blasphemers are different.

He shall be killed by being overwhelmed with stones. For he was willing to draw you away from the Lord your God, who led you away from the land of Egypt, from the house of servitude. (Deuteronomy 13:10)

When there will have been found among you, within one of your gates which the Lord your God will give to you, a man or a woman who is doing evil in the sight of the Lord your God, and who is transgressing his covenant, so as to go and serve foreign gods and adore them, such as the sun and the moon, or any of the host of heaven, which I have not instructed, and when this will have been reported to you, and, upon hearing it, if you have inquired diligently and have found it to be true, that the abomination is being done in Israel: you shall lead forward the man or the woman who has perpetrated this most wicked thing to the gates of your city, and they shall be stoned to death. (Deuteronomy 17:2–5)

And he entered, according to custom, in order to confirm the covenant, so that they would seek the Lord, the God of their fathers, with their whole heart and with their whole soul. 'But if anyone,' he said, 'will not seek the Lord, the God of Israel, let him die, from the least even to the greatest, from man even to woman.' (2 Chronicles 15:12–13)

Whoever blasphemes the name of the Lord shall be put to death. (Leviticus 24:16)

While you're at it, don't forget to kill their children, too:

Yet truly, because you have given occasion to the enemies of the Lord to blaspheme, because of this word, the son who was born to you: dying he shall die. (2 Samuel 12:14)

Blasphemy was an Early Christian issue, but now it is viewed as a major concern primarily in Islamic-majority countries. Few non-Muslim countries have laws relating to blasphemy but in Saudi Arabia, it can be a capital offense.

What if you're not that keen on killing? How should you treat those that do not rationally agree with you and with the Bible?

Now I beseech you, brethren, mark them which cause divisions and offences contrary to the doctrine which ye have learned; and avoid them. (Romans 16:17)

If there come any unto you, and bring not this doctrine, receive him not into your house, neither bid him God speed. (2 John 1:10)

Do you answer a fool? The jury's out on that one.

Answer not a fool according to his folly, lest thou also be like unto him. (Proverbs 26:4)

Answer a fool according to his folly, lest he be wise in his own conceit. (Proverbs 26:5)

Is it folly to be wise or not?

Wisdom is the principal thing; therefore get wisdom: and with all thy getting get understanding. (Proverbs 4:7)

For in much wisdom is much grief: and he that increaseth knowledge increaseth sorrow. (Ecclesiastes 1:18)

For it is written, I will destroy the wisdom of the wise, and will bring to nothing the understanding of the prudent. (1 Corinthians 1:19)

How about the righteous? Should they live?

The righteous shall flourish like the palm tree. . . . (Psalm 92:19)

The righteous perisheth, and no man layeth it to heart. . . . (Isaiah 57:1)

It's understandable that material in the Bible is incompatible with science, which evolves with each new discovery. Genesis, which is thousands of years old, totally misses the mark with regard to astronomy, physics, mathematics, and biology, not to mention geology, where the fossil evidence doesn't reconcile with Noah's ark and the flood. But why are there so many repeats and other contradictions in the Bible? What is their origin? How do they occur?

There are a number of conjectures about how this happened:

1. God couldn't find a good editor.

2. The divine inspiration process was faulty. It's hard to tell if a dream is a godly inspiration or the result of indigestion.
3. Transcribing a Bible was such a low-level task that scribes couldn't keep track of what verse they were on.
4. It's quite plausible that the stories being told by early preachers were ones shared with and by other popular preachers, who had themselves embellished them. These were the Christian stories that *worked*. Each bishop going to the Council of Nicaea (325 CE) brought his own versions of those stories. The main purpose of that council was to create an official, unified front by adopting the first approved Christian Bible canon and creed. Think of a split political party today having a convention to merge its positions/planks. There simply was not enough time to merge and edit the various stories, but they all had to be included somewhere for the group to achieve agreement.

Remember: Even a well-written fiction novel has to make sense.

In the next part we will examine the Catholic Church's partial solutions to these many Bible issues.

Part V

Catholicism

We are all born like Catholics . . . in limbo, without religion.
(Yann Martel)

Christianity has some 2.2 billion members, of which half are Roman Catholic. In this section we look more closely at Roman Catholicism, which traces its origins back to Peter and Christ and views itself as the one true church, the original from which all others split off. The Vatican City State is also recognized as the smallest independent sovereign state in the world. All that and its strong influence in Latin America and Africa makes the Catholic Church, its distinctive doctrines, and its religious, political, and social actions worthy of closer examination.

The Church as an Institution

As Eve was formed from the sleeping Adam's side, so the Church was born from the pierced heart of Christ hanging dead on the cross. (CCC 766, St. Ambrose)

The Tradition here in question comes from the apostles and hands on what they received from Jesus' teaching and example and what they learned from the Holy Spirit. The first generation of Christians did not yet have a written New Testament, and the New Testament itself demonstrates the process of living Tradition. (CCC 83)

This bears repeating:

The first generation of Christians did not yet have a written New Testament.

As any enterprise grows, it develops its culture, its credos, and its principles. In the case of a religion, those are generally embodied in its holy book. It is certainly the case that as Christianity, specifically Catholicism, developed, so did the Bible. We looked at some of its ambiguities and contradictions in the previous section. Let's look here at how the Bible was shaped and refined over the centuries and then at the doctrines that developed with it.

Bible Issues

The Catholic Church knew early on that it had a *Bible problem*. The first official New Testament Canon of Sacred Scriptures (the New Testament of the Bible), created in 325 CE by the Council of Nicaea, was badly written (in Latin) and filled with contradictions. When unauthorized translations began to be created as well, they gave people the opportunity to discover for themselves just how confounding its contents could be, something that threatened their faith and, more important possibly, the influence of the church. The church much preferred to have its priests be in control of access to that material and its interpretation.

Solution? Prohibit anyone from having or reading any translation of the Old or New Testament. That's what the Council of Toulouse (1229 CE) declared in Canon 14. The council also set up a special ecclesiastical tribunal, known as the Inquisition, to identify and deal with heretics. One option (included in Canon 6) was to destroy the house of anyone believed to be a heretic. Several years later, in 1234 CE, the Council of Tarragona ruled that Bibles written in the Romance languages must be turned in to the local bishop so they could be burned.

This worked for a few centuries. Then came Gutenberg's first printing press in 1440 and the movable type printing of the first Bible in 1455. Before this all Bibles had been copied and recopied by hand. In the 1500s Protestants (Wycliffe, Tyndale, Coverdale, et al.) began printing and distributing many translations in the vernacular. In 1559 the Council of Trent created the *Index Librorum Prohibitorum*, the list of forbidden books, and included vernacular Bibles on it. Included on the list were books by Plato, Aristotle, Cicero, Virgil, Homer, Euclid,

Hippocrates, Descartes, Locke, Hume, Hobbs, Luther, Calvin, Swift, Rousseau, Voltaire, Pascal, Kant, Locke, Kepler, Mill, Copernicus, and Galileo. Notable authors not included in later years were Darwin, Marx and Freud. Later, Hitler's *Mein Kampf* was considered for inclusion but ultimately not listed. The final edition of the Index appeared in 1948. On June 14, 1966, it was formally abolished by Pope Paul VI.

The Council of Trent also declared that officially the Catholic Church was not dependent only on the Bible, that the church could use *church tradition* in addition to the Bible in determining Catholic dogma. The Bible and church tradition would be equally valid.

And then, to really lock up its position, the council declared that the Catholic Church was the ultimate interpreter of Sacred Scripture. This remains true today and is reinforced in the present day Catholic catechism:

The task of giving an authentic interpretation of the Word of God, whether in its written form or in the form of Tradition, has been entrusted to the living teaching office of the Church alone. (CCC 85)

Should you be tempted to use reason or your conscience to oppose the church's interpretation, don't bother:

Personal conscience and reason should not be set in opposition to the moral law or the Magisterium of the Church. (CCC 2039)

For the Catholic Church, *Sacred Tradition and Sacred Scripture make up a single sacred deposit of the Word of God.* (CCC 97)

Both Scripture and Tradition must be accepted and honored with equal sentiments of devotion and reverence. (CCC 82)

This did not put an end to vernacular Bibles, of course. Pope Leo XII (May 5, 1824), Pope Pius VIII (May 24, 1829), and Pope Gregory XVI (May 8, 1844) all had occasion to lament the publication of unauthorized Bibles. In 1866, Pope Pius IX criticized Bible Societies as pests:

The crafty enemies of the Church and human society attempt to seduce the people in many ways. One of their chief methods is the misuse of the new technique of book-production. (*The Catholic Reader*, June 8, 2013)

Book production is crafty and seductive? [Wow! Thanks—Author]

This attitude has now changed. Over a span of only 800 years the church has evolved. Pope Pius XII's 1943 Papal encyclical *Divino Afflante Spiritu* endorsed reading the Bible, and Catholics joined Protestants (who had been reading the Bible all along) in doing so. Pope Francis continues to encourage the practice.

The Pope [Francis], remarking that everyone is able to read the Word of God in his or her own language, again encouraged those present always to carry a copy of the Gospel and to read one or two passages each day. (VIS September 17, 2014)

And how do we hear the voice of God? We hear it in His Word. This is why it is important to know the Scripture, as otherwise we do not know how to respond to the deceptions of the Evil One. . . . Always keep a copy of the Gospel to hand. (Pope Francis, Vatican Information Service, VIS February 21, 2015)

It's nice to see official Catholic teaching regarding a layperson reading the Bible coming into line with something that the US Constitution has guaranteed for more than two centuries. Thanks to freedom of the press, you have the right to read anything you can get your hands on. Again, the people led; the church followed.

What is the official Catholic view of the Bible today?

Sacred Scripture is the speech of God as it is put down in writing under the breath of the Holy Spirit. And [Holy] Tradition transmits in its entirety the Word of God which has been entrusted to the apostles by Christ the Lord and the Holy Spirit. It transmits it to the successors of the apostles so that, enlightened by the Spirit of truth, they may faithfully preserve, expound and spread it abroad by their preaching. (CCC 81)

The Church accepts and venerates as inspired the 46 books of the Old Testament and the 27 books of the New. (CCC 138)

For, of course, all that has been said about the manner of interpreting Scripture is ultimately subject to the judgment of the Church which exercises the divinely conferred commission and ministry of watching over and interpreting the Word of God. (CCC 119)

How do we know the pope, the bishops and teachers in the church have gotten everything right? The church explains:

The Roman Pontiff and the bishops are 'authentic teachers, that is, teachers endowed with the authority of Christ, who preach the faith to the people entrusted to them, the faith to be believed and put into practice, . . .' (CCC 2034)

And besides:

'The Roman Pontiff, head of the college of bishops, enjoys this infallibility in virtue of his office, when, as supreme pastor and teacher of all the faithful—who confirms his brethren in the faith he proclaims by a definitive act a doctrine pertaining to faith or morals . . . . The infallibility promised to the Church is also present in the body of bishops when, together with Peter's successor, they exercise the supreme Magisterium,' above all in an Ecumenical Council. When the church through its supreme Magisterium proposes a doctrine 'for belief as being divinely revealed,' and as the teaching of Christ, the definitions 'must be adhered to with the obedience of faith.' This infallibility extends as far as the deposit of divine Revelation itself. (CCC 891)

The Church . . . is held, as a matter of faith, to be unfailingly holy. (CCC 823)

In summary:

It is clear therefore that, in the supremely wise arrangement of God, sacred Tradition, Sacred Scripture and the Magisterium of the Church are so connected and associated that one of them cannot stand without the others. Working together, each in its own way, under the action of the one Holy Spirit, they all contribute effectively to the salvation of souls. (CCC 95)

The task of interpreting the Word of God authentically has been entrusted solely to the Magisterium of the Church, that is, to the Pope and to the bishops in communion with him. (CCC 100)

But isn't the church just another enterprise?

Why is the Church more than an institution? The Church is more than an institution because she is a mystery that is simultaneously human and divine. (YOUCAT 124)

Ah, that explains it: the church is a mystery.

Are God and the church the same thing?

The Church is God's presence among us men. That is why we must love her. (YOUCAT 121)

So where the church is, so also is God there? That must mean God was present and involved intimately with the Catholic Church not only during some of the nicer things it has accomplished, but also during the Crusades, the Inquisition, African slave trade, the episode with Galileo, its silence about the atrocities committed by Hitler, and the rape and torture of children in Catholic institutions? That's . . . unfortunate.

Is the Catholic Church necessary for salvation? Yes. In fact, none of the other religions can save you:

The Church is, accordingly, a sheepfold, *the sole and necessary gateway to which is Christ.* (CCC 754)

The Church exists because God wants to redeem us *together* rather than as individuals.

Why does God want there to be a Church? God wills the Church because he wants to redeem us, not individually, but together. (YOUCAT 122)

The question, *Why together?* isn't answered.

Is it true that God permitted *fallen angels* and *man's original sin* in order to show how much he loves us? Yes. Of course.

The Church is the goal of all things, and God permitted such painful upheavals as the angels' fall and man's sin only as occasions and means for displaying all the power of his arm and the whole measure of the love he wanted to give the world. (CCC 760)

This is stunning, but worth repeating. God permitted *fallen angels* and *original sin* as a means of displaying the whole measure of the love he wanted to give the world. Again, the end justifies the means. How does God escape responsibility for creating something that he knew would send most humans to perdition?

If the church's existence and mission is still not clear to you, don't worry. It is simply another mystery to accept on faith.

The Church is both visible and spiritual, a hierarchical society and the Mystical Body of Christ. She is one, yet formed of two components, human and divine. That is her mystery, which only faith can accept. (CCC 779)

Does this seem like accepting on faith the *mystery* of flying reindeer?

One God? Two? Three? Three in One?

It took seven ecumenical councils to decide exactly who Jesus was. Other councils wrestled with the question of Mary's special place in Catholicism and of the infallibility of popes.

The mystery of the Most Holy Trinity is the central mystery of the Christian faith and of Christian life. God alone can make it known to us by revealing himself as Father, Son and Holy Spirit. (CCC 261)

The Council of Nicaea (325 CE), convened by Roman Emperor Constantine the Great, a secular authority, was the first Ecumenical Council of the Catholic Church. It decided the Arian controversy within the church (was Jesus just human or human and divine?) and declared that the Son of God, Jesus, was a real God, *a true God from a true God.* It also set the Easter date formula (today it is the first Sunday after the first full moon after the first day of spring). The Nicene Creed it devised gave the various early Christian communities a basic uniform doctrine. Up to this time bishops pretty much each ran their own show. Popes with power did not exist.

The Council of Constantinople I (381 CE) was convened by Theodosius I, a Roman emperor.

This council inserted words into the Nicene Creed declaring that the Holy Spirit proceeded from both the Father and the Son. (Catholic Online)

In the third and fourth centuries the revealed truths of faith were theologically formulated and transmitted as our nonnegotiable inheritance. (Pope Francis, in *On Heaven and Earth*)

The year before the council, in 380 CE, Theodosius I had enacted a law making Christianity the official religion of the Roman Empire. This gave the Roman bishop more influence, thus he became the center of Catholic power. In about the 6th century the Roman bishop became even more influential and came to be called Pope.

The Council of Ephesus (431 CE) tried to settle a *Mary mother of God* vs. *Mary mother of Jesus* split. It defined the Catholic dogma that *the Blessed Virgin is the Mother of God* and presented the teaching of the truth of one divine person in Christ. This issue was addressed most recently again on November 11, 1994, by Pope John Paul II in the Common Christological Declaration Between the Catholic Church and the Assyrian Church of the East. Officially . . . *the Catholic tradition addresses the Virgin Mary as 'the Mother of God' and also as 'the Mother of Christ.'*

So, who created the universe? God. Well, who created God? Mary. What?

But:

God is the fullness of Being and of every perfection, without origin and without end. (CCC 213)

The Fourth Council: The Council of Chalcedon (451 CE) convened by Pope Leo the Great, resolved confusion over the idea of whether there were two natures, human and divine, in the one person of Christ, but it was rejected by Eastern Churches, which split from the main body. It issued 30 disciplinary canons governing church administration and authority.

The Council of Constantinople II (553 CE) convened by Byzantine Emperor Justinian I, reaffirmed the dogmas stated by the third and forth General Councils: The *two natures* defined at Chalcedon were now clearly interpreted as two sets of attributes possessed by a single person, Christ God, the second person of a consubstantial Trinity, i.e., one in which all are of the same substance or essence.

The Trinity is One. (CCC 553)

The Council of Toledo XI (675 CE) is the basis of three elements of the Catholic catechism:

The Trinity is One. We do not confess three Gods, but one God in three persons, the 'consubstantial Trinity.' (CCC 253)

The divine persons are really distinct from one another. (CCC 254)

The divine persons are relative to one another. . . . While they are called three persons in view of their relations, we believe in one nature or substance. (CCC 255)

The Lateran Council IV (1215 CE) reaffirmed that:

The Trinity is One. (CCC 253)

The divine persons are really distinct from one another. (CCC 254)

The council declared that *Those who give credence to the teachings of heretics, as well as those who receive, defend, or patronize them, are excommunicated.* It also decreed that only priests could perform the sacrament of the Eucharist and that Jews could not hold public office.

The Council of Florence (1442 CE) declared again that *The divine persons are relative to one another.*

The Council of Trent (1545–1563 CE) ruled that Jesus is God and that the commandments are to be followed by Catholics.

We therefore hold, with the Council of Trent, that original sin is transmitted with human nature, 'by propagation, not by imitation' and that it is . . . 'proper to each. . . .' (CCC 419)

. . . Still, the transmission of Original Sin is a mystery that we cannot fully understand. But we do know by Revelation that Adam had received original holiness and justice not for himself alone, but for all human nature. By yielding to the tempter, Adam and Eve committed a personal sin, but this sin affected the human nature that they would then transmit in a fallen state. (CCC 404, Council of Trent)

Another mystery is created and then propagated.

On December 8, 1854, Pope Pius IX declared the Immaculate Conception of Mary to be official Church dogma. It holds that Mary herself was conceived without the *normal* stain of original sin and was predestined to be the mother of Christ.

Through the centuries the Church has become ever more aware that Mary, 'full of grace' through God, was redeemed from the moment of her conception. That is what the dogma of the Immaculate Conception confesses, as Pope Pius IX proclaimed in 1854. . . . (CCC 491)

Ten years later, in 1864, Pope Pius IX condemned all scientific discoveries unless approved by the church. And six years after that, Vatican I was convened. Called to deal with the threat of the rising influence of rationalism (The Age of Reason), it defined and declared the existence of *papal infallibility.* Clearly, science and reason were deemed to be real threats.

The supreme degree of participation in the authority of Christ is ensured by the charism of infallibility. This infallibility extends as far as does the deposit of divine Revelation; it also extends to all those elements of doctrine, including morals, without which the saving truths of the faith cannot be preserved, explained, or observed. (CCC 2035)

This is My Body . . . Really!

The Council of Trent in Session 13, October 11, 1551, also reaffirmed the transubstantiation—bread into body, wine into blood—of Jesus Christ, via the Catholic sacrament of the Eucharist. And no, they absolutely don't mean that symbolically.

Truly, I say to you, unless you eat the flesh of the Son of man and drink his blood, you have no life in you. (CCC 1384, John 6:53, quoting Jesus)

I*n the most blessed sacrament of the Eucharist 'the body and blood, together with the soul and divinity, of our Lord Jesus Christ and, therefore, the whole Christ is truly, really, and substantially contained.'* (CCC 1374, Council of Trent)

This means, yes, that when you take Communion, you really and truly eat Jesus and drink his blood. The catechism is so anxious for you to understand the importance of the Eucharist that it says it again and again:

The Eucharist is the center of the life of the particular Church. . . . (CCC 893)

The Eucharist is the source and summit of the Christian life. (CCC 1324)

The Eucharist makes the Church. (CCC 1396)

The Eucharist is the heart and the summit of the Church's life . . . (CCC 1407)

The Eucharist is based on the following Bible verses:

And as they were eating, Jesus took bread, and blessed it, and brake it, and gave it to the disciples, and said, Take, eat; this is my body. And he took the cup, and gave thanks, and gave it to them, saying, Drink ye all of it; For this is my blood of the new testament, which is shed for many for the remission of sins. (Matthew 26:26-28)

Catholic Mass with its Eucharist *celebration* is also based on Paul's 1 Corinthians 11:23–25.

This, the oldest account of the events in the Upper Room at the Last Supper, is by the APOSTLE Paul, who was not an eyewitness himself, but rather wrote down what was being preserved as a holy mystery by the young Christian community and was being celebrated in the liturgy. (YOUCAT 210)

In brief, the Eucharist is the sum and summary of our faith: 'Our way of thinking is attuned to the Eucharist, and the Eucharist in turn confirms our way of thinking.' (CCC 1 327)

In the Eucharist Christ gives us the very body which he gave up for us on the cross, the very blood which he 'poured out for many for the forgiveness of sins.' (CCC 1365)

The principal fruit of receiving the Eucharist in Holy Communion is an intimate union with Christ Jesus. (CCC 1391)

To get intimate with Jesus, drink his blood. Sounds positively Goth, doesn't it?

By the same charity that it enkindles in us, the Eucharist preserves us from future mortal sins. (CCC 1395)

So, if you eat Jesus's body and drink Jesus's blood you won't kill anyone or rob a bank?

In the Eucharist, the Church expresses her efficacious communion with the departed. . . . (CCC 1689)

That is, your drinking Jesus's blood helps the church communicate with the dead.

The Church warmly recommends that the faithful receive Holy Communion when they participate in the celebration of the Eucharist; she obliges them to do so at least once a year. (CCC 1417)

The Sunday celebration of the Lord's Day and his Eucharist is at the heart of the Church's life. 'Sunday is the day on which the paschal mystery [life, passion, death, resurrection and ascension of Jesus Christ] is celebrated in light of the apostolic tradition and is to be observed as the foremost holy day of obligation in the universal Church.' (CCC 2177, CCL 1246)

The Sunday Eucharist is the foundation and confirmation of all Christian practice. (CCC 2181)

Supernatural *Extras*

Catholicism also has Heaven, Hell, exorcism, a virgin mother, and rising from the dead. And if that's not enough hocus-pocus for you, the church throws in a number of supernatural beings.

The existence of the spiritual, non-corporeal beings that Sacred Scripture usually calls 'angels' is a truth of faith. The witness of Scripture is as clear as the unanimity of Tradition. (CCC 328)

The Church venerates the angels who help her on her earthly pilgrimage and protect every human being. (CCC 352)

Catholicism isn't as monotheistic as most think. It has its hierarchy of Triune God, devil, angels, and saints, all with various god powers and special responsibilities.

Devil—omnipresent, but not omnibenevolent and not omnipotent

Angels—omnipresent, but not omnipotent and not omnibenevolent (Remember—the devil was a fallen angel.)

Mary—omnipresent and influential, but not omnipotent

Saints—omnipresent and influential, but not omnipotent

Saints are like Santa (aka Saint Nicholas)—omnipresent and omniscient. We don't call Santa omnibenevolent—because he can treat children badly (e.g., withhold toys or leave pieces of coal). Oddly enough, though, we still like to consider God as omnibenevolent even though he can send people to a perpetual suffering for reasons he devised, in a hell he created.

Believers often pray to the church's saints, which number more than 10,000, as well as wearing medals and other talismans that are supposed to protect the wearers from harm of various sorts.

We can and should ask them [Saints] to intercede for us and for the whole world. (CCC 2683)

But the Bible says: *For there is one God, and one mediator also between God and men, the man Christ Jesus.* (1 Timothy 2:5)

This doesn't leave much room for angels, Mary, and the saints as intercessors. Then again, why should someone pray to anyone other than God? Do they think this will be more effective? That the saints are less busy than God or more likely to influence him than they are? Being omniscient, God should recognize your needs long before your favorite saint is made aware of them. And wouldn't claimed success of saintly intercessions reflect poorly on the claimed omniscience and benevolence of God?

The question of the effectiveness of prayers to saints is probably as helpful as arguing about the number of angels that can dance on a pin. I'm guessing, in fact, that saintly intercessions to God succeed about as often as our intercessory prayers to God succeed.

Justification: Jesus Saves—By Faith Alone or by Faith and Good Works?

Faith is necessary for salvation. The Lord himself affirms: 'He who believes and is baptized will be saved; but he who does not believe will be condemned.' (Mark 16:16, CCC 183)

Believing in Jesus Christ and in the One who sent him for our salvation is necessary for obtaining that salvation. (CCC 161)

Wherefore the law was our schoolmaster to bring us unto Christ, that we might be justified by faith. (Galatians 3:24)

Ye see then how that by works a man is justified, and not by faith only. (James 2:24)

Although it is grace and faith through which we are saved, nevertheless, our good works aught to show the love produced by God's action in us. (YOUCAT 341)

Therefore we conclude that a man is justified [saved] by faith apart from the deeds of the law. (Romans 3:28)

Therefore being justified by faith, we have peace with God through our Lord Jesus Christ. (Romans 5:1)

For by grace are ye saved through faith; and that not of yourselves: it is the gift of God: Not of works, lest any man should boast. (Ephesians 2:8-9)

Even so faith, if it hath not works, is dead, being alone. (James 2:17)

Knowing that a man is not justified by the works of the law but by faith in Jesus Christ, even we have believed in Christ Jesus, that we might be justified by faith in Christ and not by the works of the law; for by the works of the law no flesh shall be justified. (Galatians 2:16)

Note that good works alone cannot save you. You must have faith in Jesus.

I invite reflection...on the connection between faith and truth, ... Faith knows because it is tied to love, because love itself brings enlightenment. Faith's understanding is born when we receive the immense love of God which transforms us inwardly and enables us to see reality with new eyes. (Pope Francis January 28, 2014)

Maybe you didn't know this: A good way to get into heaven is to have yourself castrated. *For there are eunuchs who were born that way from their mother's womb; and there are eunuchs who were made eunuchs by men; and there are also eunuchs who made themselves eunuchs for the sake of the kingdom of heaven. He who is able to accept this, let him accept it.* (Matthew 19:12)

Finally, Pope Francis turned to the forgiveness of sins. In the Sacrament of Baptism 'all sins are forgiven, original sin and all personal sins, as well as all forms of punishment for sin.' (VIS November 13, 2013)

Baptism is a fresh start. Perhaps baptizing someone right after birth should be replaced with baptism just before death? You might respond that we don't know when we will die. What about those that accept euthanasia?

The mission of the Church, following the mandate of the Risen Christ is to evangelise [sic] and to forgive the sins through Sacramental Baptism. (Pope Francis, VIS November 13, 2013)

Summarizing the Biblical and Catholic proclamations regarding justification, being *saved*:

Can someone earn heaven by good works? No. No man can gain heaven merely by his own efforts. (YOUCAT 341)

So, you can't get into heaven solely through your own efforts—but before you meet St. Peter at the Pearly Gates, here are some *dos* and *don'ts* to check off on your *Go to Heaven* list:

DO . . .

Have faith in the Jesus story—Jesus is God, was crucified, rose from the dead, etc. (See John 14:6)

Good works (See Matthew 16:27)

Obey the Ten Commandments (See Council of Trent, 1569, CCC 2068)

Get baptized (See Council of Trent, 1569, CCC 1257)

Be a Catholic (See CCC 754)

Accept a sufficient amount of God's Grace (See Council of Trent)

Seek the truth and follow your conscience (See CCC 1260)

Be among the 144,000 declared chosen by God (See Revelation 7:1-10)

Publicly proclaim that Jesus rose from the dead (See Romans 10:9)

Abandon family ties (See Matthew 19:29)

Pray (See CCC 2744)

DO NOT . . .

Die with an unforgiven mortal sin in your background (See CCC 1035)

Criticize a priest or bishop or the church (See CCT)

Use rational thought to argue with the church (See CCC 2039)

Have your *privy member* cut off (See Deuteronomy 23:1)

Be rich, have assets (See Matthew 19:24)

Have bastards in your family for the previous ten generations: *A bastard shall not enter into the congregation of the Lord; even to his tenth generation shall he not enter into the congregation of the Lord.* (See Deuteronomy 23:2)

Note: Just to be clear on that last one—If any one of your 2,046 relatives (your parents, their parents, etc.) for your previous ten generations were born of parents that were not married, you're screwed and can't get into heaven no matter what you believe or do. This is especially noteworthy since Gallup reports (May 22, 2012) that 54 percent of Americans now believe having a baby outside of marriage is morally acceptable. Is this why people do genealogy? You'll notice how they carefully track births and weddings. They're wondering if they even have a chance?

Should any of the above prove to be a stumbling block, you may be out of luck—unless you know a Mormon who will arrange for a posthumous proxy baptism for you as the Mormon Church did for Anne Frank and other victims of the Holocaust and for Ann Romney's atheist father.

Heaven is only temporary anyway. So why bother?

Heaven and earth shall pass away, but my words shall not pass away. (Matthew 24:35)

I am curious about what happens to those already in heaven when it passes away. Poor things.

But, not so fast: *One generation passeth away, and another generation cometh: but the earth abideth for ever.* (Ecclesiastes 1:4)

Heaven is temporary but earth is forever?

And finally—Catholic Canon 989, *After having reached the age of discretion, each member of the faithful is obliged to confess faithfully, his or her grave sin[s] at least once a year.*

Both *age of discretion* and *grave sin* are rather subjective. What if my understanding differs from God's? I'll never know, till I die.

But what about the naïve? Will people who have never heard of Jesus go to hell? This is a most difficult question for many Christians.

Believing in Jesus Christ and in the One who sent him for our salvation is necessary for obtaining that salvation. (CCC 161)

The Lord himself affirms that Baptism is necessary for salvation. . . . The Church does not know of any means other than Baptism that assures entry into eternal beatitude. . . . (CCC 1257)

Except . . . a page later in the same catechism the church finds a way:

Every man who is ignorant of the Gospel of Christ and of his Church, but seeks the truth and does the will of God in accordance with his understanding of it, can be saved. (CCC 1260)

Baptism is the only way to God and salvation; . . . In places where the Church does not exist or has had no success—whether by her own fault or for other reasons—God himself paves for the people other ways to salvation in Christ. (YOUCAT 199)

But these other ways are God's secret.

Where do souls go after death if not to heaven?

Immediately after death the souls of those who die in a state of mortal sin descend into hell, where they suffer the punishments of hell, 'eternal fire.' (CCC 1035)

A cruel and unusual punishment indeed. Our government, in accordance with the Eighth Amendment to the Constitution, would not allow this punishment. And God made the mortal sin/hell rule. Therefore God is meaner than the United States government. Religion still lags the civil society on this one.

Happily, the church has offered an *out* for lesser transgressors—those souls who have died who don't merit hell but aren't quite ready for heaven end up in purgatory.

All who die in God's grace and friendship, but still imperfectly purified, are indeed assured of their eternal salvation; but after death they undergo purification, so as to achieve the holiness necessary to enter the joy of heaven. The Church gives the name Purgatory to this final purification of the elect, which is entirely different from the punishment of the damned. (CCC 1030-1031)

Their sins have been forgiven by God, but atonement for them still must be made, and these souls must be purged of any remaining attachments to sin or other such imperfections before entering paradise. This is accomplished by their loved ones praying for them. It was also the basis of the selling of *indulgences* in the late medieval period that were part of what motivated Martin Luther to present his *95 Theses.*

Faith: The Essence of the Church and Christianity

What moves us to believe is not the fact that revealed truths appear as true and intelligible in the light of our natural reason; we believe because of the authority of God himself who reveals them, who can neither deceive nor be deceived. . . .' (CCC 156)

During the Reformation, Protestants promoted the belief that every believer was a priest and could interpret the Bible for his or her self. Catholics continued to promote the mystery of faith. Let's see what the Catholic Church's official catechism has to say about faith today.

It is right and just to entrust oneself wholly to God and to believe absolutely what he says. It would be futile and false to place such faith in a creature. (CCC 150)

We can believe in Jesus Christ because he is himself God. (CCC 151)

I and my Father are one. (John 10:30)

At the same time Jesus gives us to understand that he himself is 'the Lord.' To confess that Jesus is Lord is distinctive of Christian faith. This is not contrary to belief in the One God. (CCC 202, Mark 12:35-37)

If this is really true, then after the crucifixion the universe was without a god for about three days. It just coasted for three days? No conceptions? No deaths? No new miracles?

Faith is a gift of God, a supernatural virtue infused by him. (CCC 153)

Faith is not only good, but supernatural?

The most common yet most hidden temptation is our lack of faith. (CCC 2732)

Trusting in God and cleaving to the truths he has revealed is contrary neither to human freedom nor to human reason. (CCC 154)

Faith is certain. It is more certain than all human knowledge because it is founded on the very word of God who cannot lie. (CCC 157)

This does not quite explain why the church has trouble reconciling its faith with the findings of science. But still the church claims it works:

Because of the fundamental questions it has always raised, the exploration of the universe has also deepened the understanding of faith and its rapport with science. The Holy See believes that faith is capable of both expanding and enriching the horizons of reason. (VIS, 3 November 2014)

The grace of faith opens 'the eyes of your hearts' to a lively understanding of the contents of Revelation. . . . (CCC 158)

Faith is the theological virtue by which we believe in God and believe all that he has said and revealed to us, and that Holy Church proposes for our belief, because he is truth itself. (CCC 1814)

To better reinforce faith, the believers should do the following:

To nourish the faith of believers, the signs which accompany the Word of God should be emphasized: the book of the Word (a lectionary or a book of the Gospels), its veneration (procession, incense, candles), the place of its proclamation (lectern or ambo), its audible and intelligible reading, the minister's homily which extends its proclamation, and the responses of the assembly (acclamations, meditation psalms, litanies, and profession of faith). (CCC 1154)

We should always be disposed to believe that that which appears white is really black, if the hierarchy of the Church so decides. (Ignatius Loyola, First Superior General of the Jesuit Order, the early religious home of Pope Francis)

The preceding quote could almost be humorous, if it wasn't so weird.

Can you be a Christian without believing in the Resurrection of Christ? No. If Christ has not been raised, then our preaching is in vain and your faith is in vain. (1 Corinthians 15:14). (YOUCAT 104)

In a part of the Catholic Catechism called *The Profession of Faith* it states: *Whoever says 'I believe' says 'I pledge myself to what we believe.'* (CCC 185)

This is an excellent example of theological thinking. The church simply states something it wishes were true and, voilà, it is true. One really can't believe anything by oneself. One can only believe what *we* believe.

To help you learn and believe in the resurrection of Jesus, it is celebrated every Sunday during Mass.

We must learn on earth how to celebrate properly, or else we will not know what to do in heaven. (YOUCAT 187)

Just imagine! An unlearned human makes it all the way to heaven only to become a sad outcast for never having learned how to celebrate.

To be truly faithful, you also have to learn how to submit to the will of God—or at least of his church.

By faith, man completely submits his intellect and his will to God. With his whole being man gives his assent to God the revealer. Sacred Scripture calls this human response to God, the author of revelation, 'the obedience of faith.' (CCC 143)

To obey in faith is to submit freely to the word that has been heard, because its truth is guaranteed by God, who is Truth itself. (CCC 144)

In faith, the human intellect and will cooperate with divine grace: 'Believing is an act of the intellect assenting to the divine truth by command of the will moved by God through grace.' (CCC 155, St. Thomas Aquinas)

It's not reasonable, but God's authority somehow compels us to believe. You don't have to worry about God deceiving you.

Faith is certain. It is more certain than all human knowledge because it is founded on the very word of God who cannot lie. To be sure, revealed truths can seem obscure to human reason and experience, but 'the certainty that the divine light gives is greater than that which the light of natural reason gives.' (CCC 157, St. Thomas Aquinas)

Divine light trumps natural reason. Faith in God is legit because God can't lie.

Ten thousand contradictions, inconsistencies, and mysteries are not sufficient to justify a doubt in one item of faith? But in reality, one inconsistency or contradiction tears down the story.

Ten thousand difficulties do not make one doubt. (CCC 157, John Henry Cardinal Newman)

What is Faith? Faith is a supernatural gift of God, which enables us to believe without doubting whatever God has revealed. (The Penny Catechism, 9)

In the words of St. Augustine, I believe, in order to understand; and I understand, the better to believe. (CCC 158)

Faith helped St. Augustine understand, and understanding promoted his faith.

The act of faith is of its very nature a free act. (CCC 160)

Man's faculties make him capable of coming to a knowledge of the existence of a personal God. But for man to be able to enter into real intimacy with him, God willed both to reveal himself to man and to give him the grace of being able to welcome this revelation in faith. The proofs of God's existence, however, can predispose one to faith and help one to see that faith is not opposed to reason. (CCC 35)

But without faith it is impossible to please him: for he that cometh to God must believe that he is, and that he is a rewarder of them that diligently seek him. (Hebrews 11:6)

Faith is everything, in fact.

In the Rituale Romanum, *the minister of Baptism asks the catechumen: 'What do you ask of God's Church?' And the answer is: 'Faith.' 'What does faith offer you?' 'Eternal life.'* (CCC 168)

The Church's faith precedes, engenders, supports and nourishes our faith. The Church is the mother of all believers . . . (CCC 181)

We believe all 'that which is contained in the word of God, written or handed down, and which the Church proposes for belief as divinely revealed.' (CCC 182)

The One True Church

Not surprisingly, in the course of hammering out these doctrines it was not unusual for those on the losing sides of the arguments to split off from the main body. One such separation, called the *Great Schism*, occurred in 1054 CE.

From the 4th century onwards five patriarchs had governed regions of the church from Rome (the Western church) and Antioch, Alexandria, Jerusalem, and Byzantium (the Eastern church). Gradually the Eastern churches grew independent of Rome and their doctrines diverged. Among other things, the Roman church required a celibate clergy and the Eastern churches allowed married clergy. Eventually the Roman and Greek popes excommunicated each other, finalizing the split. (The Second Vatican Council lifted its excommunication in 1965, almost 10 centuries later.)

During the Reformation a number of other groups split from the Roman Catholic Church to protest what they saw as its excesses and errors. In 1517, with Martin Luther at the forefront criticizing the sale of indulgences and the church's assertion of the supremacy of church authority over that of the Bible, Western Christianity split into Catholic and Protestant branches. In 1999 the Catholics and Lutherans in a joint declaration announced an agreement on the doctrine of justification. Good works would not save a person, they said, but faith and grace would. They also agreed not to condemn each other's faith.

Catholics like to claim *true lineage* back to Peter and Christ in order to prove the truth and validity of their dogma, but other branches on the Christian tree such as Lutherans, Anglicans, and Eastern Orthodox can make the same claim through apostolic succession back to the early Christian community following Christ. Imagine the Mississippi River watershed. If one places Christ at a lower delta lobe it's easy to trace with, say, a floating cork from any Christian upper tributary to the delta. It's not so easy to do the reverse—to go upstream and stay on the Mississippi, the *true* river—not unless we simply declare that the Mississippi begins at Lake Itasca. But why start there? The Missouri branches from the Mississippi and is longer, and the Ohio branch supplies more water to the overall flow. Maybe tradition does somehow trump logic. Or, each branch could have its own pope and life goes on like that.

The main question here is: How does a church determine official lineage when splits occur, something which has happened on numerous occasions over the centuries? We may get a chance to study the phenomenon in our lifetime. Anglicanism today is splitting over the *gay issue* and the Episcopal Church [the American branch of Anglican Church], accusing the Africans of poaching on its territory, *has launched a raft of lawsuits across the US to get back property occupied by dissident American congregations. The Rev. Susan Russell, president of Integrity, a group that champions gay and lesbian rights in the Episcopal*

Church, denounces offshore bishops as the 'intercontinental ballistic weapons of schism and division.' (*The Wall Street Journal*/WSJ.com, September 20, 2007)

If something similar happened to the Catholic Church, which branch would be the *real* or *true* Catholics? Both sides would be traceable back to the same early popes. I'm guessing whichever side ends up with the Sistine Chapel and the Vatican Museums after a legal tussle will be perceived as the legitimate Catholics. We'll just have to wait and watch.

The Catholic Church as a Nation-state

The Catholic Church is not just a religion. It is also a country, thanks to Benito Mussolini.

From 1922 until 1943, Benito Mussolini was prime minister of Italy, one third of the World War II era equivalent of the *Axis of Evil*, the nations that fought against the Allied forces. During this period the fascist dictator made a number of notable statements, including these:

We do not argue with those who disagree with us, we destroy them.

The history of saints is mainly the history of insane people.

The truth is that men are tired of liberty.

Fascism is a religious concept.

Liberty is a duty, not a right.

I am making superhuman efforts to educate this people. When they have learnt to obey, they will believe what I tell them.

Democracy is beautiful in theory; in practice it is a fallacy. You in America will see that some day.

It's easy to see then why Pope Pius XI said this about him in 1932: *Mussolini: a gift from providence.*

To be fair, at the time he was referring to Mussolini's role in the ratification of the Lateran Treaties in 1929. In these, Italy recognized the Vatican as an independent state and Mussolini agreed to give the church financial support. Italy gave the Catholic Church land (108 acres) and a large sum of cash. In return Mussolini received public support from the pope. The Lateran Treaty/Pact/Accords, modified in 1984, is now included within the Italian Constitution.

The Lateran Treaty of 1929 also grants the Holy See extraterritorial authority over 23 sites in Rome and five outside of Rome, including the Pontifical Palace at Castel Gandolfo (the Pope's summer residence) . . . defense is the responsibility of Italy. (US CIA)

A Sovereign Nation

Then Peter and the other apostles answered and said, we ought to obey God rather than men. (Acts 5:29)

In international law, a sovereign state is a nonphysical juridical entity that is represented by one centralized government that has sovereignty over a geographic area. International law defines sovereign states as having a permanent population, defined territory, one government, and the capacity to enter into relations with other sovereign states. (See Wikipedia)

The Holy See is the supreme authority of the Roman Catholic Church. It evolved into existence long ago as a non-territorial enterprise, but today it is a sovereign state which operates from and represents the Vatican City State, over which the Holy See exercises full territorial sovereignty. The Vatican City State is recognized as the smallest independent sovereign state in the world, and Italy sees the wall of the Vatican (within Rome) as a national border.

The Vatican City State is a divine-right absolute monarchy. The pope, the Supreme Pontiff, is the head of state with full legislative, judicial and executive powers and is the ruler of both the Vatican City State and the Holy See. As head of state, the pope travels with diplomatic immunity from arrest.

The Vatican City State issues its own stamps and coins, including a coin that is legal tender across Europe. It has a set of criminal laws. It also has its own police force, though it contracts with Italy for some related services. (When Pope Benedict's former butler received a three-year sentence for stealing documents, for instance, it would have been served in an Italian jail since the Vatican City State has no suitable facilities. (BBC News) The pope later pardoned him.)

The Holy See is a sovereign juridical entity under international law. The United States and the Holy See consult and cooperate on international issues of mutual interest, including human rights, inter-religious understanding, peace and conflict prevention, development, and environmental protection. The US has an official ambassador to the Holy See. The first US ambassador to the Holy See, William A. Wilson, was appointed in 1984 by President Reagan. The present US ambassador is Ken Hackett. The Holy See's ambassador to the US is Archbishop Carlo Maria Vigano. The Holy See maintains full diplomatic relations with 180 other states. (See Vatican City, VIS 22 March 2013)

The Holy See is a *non-member Permanent Observer State* within the U.N. (See Resolution 58/314 of July 16, 2004) It has all the rights of a member except the rights to vote and to present candidates. No other religion in the world has the status as a state and membership in the U.N. The current Holy See Permanent Observer to the United Nations is Archbishop Silvano M. Tomasi. (See VIS, March 27, 2015)

The Holy See, in ratifying the Comprehensive Nuclear Test Ban Treaty (CTBT), adopted by the United Nations General Assembly on 10 September 1996 and signed by the Holy See on 24 September of the same year, wishes to repeat what was said when it added its signature . . . (U.N. document)

Holy See Concordats for the Vatican are like Status of Forces Agreements for the United States. They are accords that define the rules of occupation within the host countries—rules like separate privileged schools, tax privileges, sovereign state privileges, moral arbiter, and access to state resources to further its own self interest. The church claims it is autonomous and independent of the local civil political community and its laws under the principle of religious freedom.

To the Church belongs the right always and everywhere to announce moral principles, including those pertaining to the social order, and to make judgments on any human affairs to the extent that they are required by the fundamental rights of the human person or the salvation of souls. (Pope John Paul II, COMPENDIUM OF THE SOCIAL DOCTRINE OF THE CHURCH)

In 1977 in Ireland the Vatican argued that Church Canon Law relating to *mandatory reporting* of suspected child-abuse cases should take precedence over Irish civil law.

The US and the Holy See signed a Memorandum of Understanding on May 7, 2013, to fight money laundering and terrorism financing globally.

Vatican and U.S. officials signed an agreement in which the Holy See committed to comply with a 2010 U.S. law designed to encourage—some say force—foreign financial institutions to share information about U.S. account holders with U.S. tax authorities. . . . But the Vatican foreign minister, Monsignor Paul Gallagher, said the Vatican's decision to sign its first intergovernmental agreement with the U.S. was part of its 'long-term strategy to ensure and promote legality, transparency and ethical behaviour in the economic and financial fields.' (AP, June 10, 2015)

Churches in the US are typically recorded as 501(c)(3) non-profits, but they are not required to file Form 990 disclosure forms and are tax-free. Churches are exempt from most civil regulations—anti-discrimination laws, education content and goals, and the Americans with Disabilities Act. Churches are not required to file a tax form.

Catholic Dogma: The Big Questions

Catholic dogma originates within a foreign Holy See/Vatican City State or the supernatural world.

What faith assures us of is true even in the civil sphere. . . .
(Pope Francis, June 8, 2013)

Better for us that you read that again. Incredible.

But our citizenship is in heaven, and it is from there that we are expecting a Savior, the Lord Jesus Christ. (Philippians 3:15-21)

Each and every thing which is proposed definitively by the magisterium of the Church concerning the doctrine of faith and morals, that is, each and every thing which is required to safeguard reverently and to expound faithfully the same deposit of faith, is also to be firmly embraced and retained; therefore, one who rejects those propositions which are to be held definitively is opposed to the doctrine of the Catholic Church. (CCL 750, §2)

No one may follow orders from the State that violate God's Laws. (YOUCAT 377)

We must obey God rather than men. (Acts 5:29)

Citizens are obliged in conscience not to follow the directives of civil authorities when they are contrary to the demands of the moral order. *We must obey God rather than men.* (CCC 2256, Acts 5:29)

We are Catholics before we are Americans. (Archbishop Charles J. Chaput, 1-23-2012)

The political duty of Catholics . . . is to be Catholic first . . . and act like faithful Catholics all the time. (Archbishop Charles Chaput)

Let every soul be subject unto the higher powers. For there is no power but of God: the powers that be are ordained of God. Whosoever therefore resisteth the power, resisteth the ordinance of God: and they that resist shall receive to themselves damnation. (Romans 13:1-2)

So, God ordained Hitler, Stalin, Mussolini, Osama Bin Laden, Pol Pot and Kim Jong Un? And the Bible says, *they that resist shall receive to themselves damnation.* You might find this hard to believe.

The Church supports democracy because of all political systems it offers the best conditions for achieving equality before the law and safeguarding human rights. (YOUCAT 441)

More so than a Catholic monarchy?

Democracy operates on the principle that all power comes from the people. In the Church, however, all power comes from Christ. (YOUCAT 140)

But even in the Church, power comes from the top while influence and money comes from the bottom. The Bible contradicts the Constitution—they offer two extremes: God power vs. people power. In the United States the *we the people* constitution took the ultimate power from the church.

The Most High holds power over the kingdom of men, and that whoever he wishes, he will set over it. (Daniel 5:21)

Catholic Canon Law

Civil laws to which the law of the Church yields are to be observed in canon law with the same effects, insofar as they are not contrary to divine law and unless canon law provides otherwise. (CCL 22)

In the view of the Catholic Church, Canon Law prevails over civil law.

The Church has the innate and proper right to coerce offending members of the Christian faithful with penal sanctions. (CCL 1311)

Again: The Catholic Church believes it can send a Christian to jail.

The law can establish other expiatory penalties which deprive a member of the Christian faithful of some spiritual or temporal good and which are consistent with the supernatural purpose of the Church. (CCL 1311-2)

Although not an assent of faith, a religious submission of the intellect and will must be given to a doctrine which the Supreme Pontiff or the college of bishops declares concerning faith or morals when they exercise the authentic magisterium, even if they do not intend to proclaim it by definitive act; therefore, the Christian faithful are to take care to avoid those things which do not agree with it. (CCL 752)

A person must believe with divine and Catholic faith all those things contained in the word of God, written or handed on, that is, in the one deposit of faith entrusted to the Church, and at the same time proposed as divinely revealed either by the solemn magisterium of the Church or by its ordinary and universal magisterium which is manifested by the common adherence of the Christian faithful under the leadership of the sacred magisterium; therefore all are bound to avoid any doctrines whatsoever contrary to them. (CCL 750, §1)

Pronouncements of this kind were certainly something that concerned certain voters in 1960 (the fact that John Kennedy was Catholic raised worries that he would be in thrall to the pope). Times have changed, but statements in the Catholic catechism and by the pope suggest what some of the ramifications of strict adherence to this doctrine might be.

Professional secrets—for example, those of political office holders,

soldiers**, physicians, and lawyers—or confidential information given under the seal of secrecy must be kept, save in exceptional cases where keeping the secret is bound to cause very grave harm to the one who confided it, to the one who received it or to a third party and where the very grave harm can be avoided*** *[emphasis added], only by divulging the truth. Even if not confided under the seal of secrecy, private information prejudicial to another is not to be divulged without a grave and proportionate reason.* (CCC 2491)

The Archdiocese for the Military Services (AMS) provides the Roman Catholic Church's pastoral and spiritual services to those serving in the United States armed forces or other federal services overseas. It oversees Catholic priests serving as military chaplains throughout the world. *In 2013 the AMS served at more than 220 U.S military installations in 29 countries.* (AMS) Does the above mean, for instance, that those chaplains have the choice of passing on political or military secrets to a third party . . . like the Vatican?

Imagine what it would be like if the Vatican collected information from everywhere and ran everything. Pope Benedict XVI did:

To manage the global economy; to revive economies hit by the crisis; to avoid any deterioration of the present crisis and the greater imbalances that would result; to bring about integral and timely disarmament, food security and peace; to guarantee the protection of the environment and to regulate migration: for all this, there is urgent need of a true ***world political authority*** *[emphasis added], as my predecessor Blessed John XXIII indicated some years ago. Such an authority would need to be regulated by law, to observe consistently the principles of subsidiarity and solidarity, to seek to establish the common good, and to make a commitment to securing authentic integral human development inspired by the values of charity in truth.* (Pope Benedict XVI, December 3, 2012)

There are already too many examples of instances where the church regarded itself as a *true world political authority* and a law unto itself.

Archbishop Jozef Wesolowski, former nuncio (ambassador) to the Dominican Republic, was holed up in Poland where the Vatican rejected a request from the Warsaw Prosecutor's Office for his extradition on the grounds he was a citizen of the Vatican. (*RNS*) More recently, Wesolowski reportedly has been defrocked and has lost his diplomatic immunity. *The Vatican's former ambassador to the Dominican Republic has been charged with sexually abusing young boys in the Caribbean country and having child pornography on his computer and will stand trial next month in a Vatican court.* (VATICAN CITY, AP, June 2015)

In St. Paul, Minnesota, Archbishop [Joseph] Flynn did not report any charges of clergy sex abuse to police over a thirteen-year period. The archdiocese did finally agree to a 'child protection plan' that requires the archdiocese report any child abuse claim to law enforcement and give law enforcement priority in any investigation. (*StarTribune*, October 14, 2014)

In a court-ordered deposition made public Thursday, the Rev. Kevin McDonough [former vicar general for the Archdiocese of St. Paul and Minneapolis] also said priests suspected of abusing children were not automatically reported to police, as required by law. (*StarTribune*, April 25, 2014)

Maybe the pope should declare that all bishops and priests are *diplomats* and thus gain diplomatic immunity for them from US criminal laws. That worked (for a while) for one archbishop.

In Australia, papal envoy Archbishop Paul Gallagher invoked diplomatic immunity in refusing to provide documents that could have assisted a sex abuse inquiry. (*New York Times*, May 23, 2014)

The Lobbying Disclosure Act, passed by Congress in 1995, provides exemptions for churches or an association of churches. So, a [Catholic/Muslim/Religious] foreign state not only has total freedom to educate American children as they wish in their parochial schools and churches, with no oversight, but in addition, the foreign state can lobby our government to their benefit without registering or disclosing the lobbying effort. (American citizens can't even do that.) And they do it tax free, as a religion. (Note: It is otherwise a crime to be an unregistered agent of a foreign state.)

Unlike Secular lobbies, religious groups that work with legislators on Capitol Hill don't have to register with the federal government and are free from the stringent reporting requirements imposed on any group that seeks to influence legislation. (Church and State, April 2012)

The Fort Wayne-South Bend Catholic Diocese in Indiana has refused to appear in court to answer a lawsuit from a former employee, arguing that it is beyond the reach of the legal system in this case. (Church and State, January 2015)

In effect until 1983, Church Canons 119-120 called for criminally accused clerics to be tried before a church judge and explicitly forbade taking to civil court a cleric accused of a crime. (*StarTribune*, Ed Kohler, St. Paul, January 2015)

The Archdiocese in St. Louis recently excommunicated the local priest and six local St. Stan parishioners because they would not relinquish the deed to their church to the bishop. The locals had built, owned, and operated the Catholic parish church since the 1890s. The Catholic hierarchy obviously valued a church asset more than the souls of the legal owners.

The bishop in Spokane argued in the opposite direction in its bankruptcy suit, claiming even though the diocese had the deeds, that the court cannot use them as liquid assets. The US bankruptcy judge disagreed. In the United States, civil law should prevail over canon law.

As noted earlier, times have changed . . . and so have the attitudes of the church toward civil society. Take this statement from the *Baltimore Catechism*, the de facto standard Catholic school text in the United States from 1885 to the late 1960s, for instance.

Why must we respect and obey the lawful authority of our country? We must respect and obey the lawful authority of our country because it comes from God, the Source of all authority. (Baltimore Catechism, 247)

Mind you, Pope Pius XII had a different take on that:

Above all, it must be clearly stated that no human authority, no state, no community of states, whatever be their religious character, can give a positive command or positive authorization to teach or to do that which would be contrary to religious truth or moral good. (Pope Pius XII, Dec. 6, 1953)

What does the current catechism say about such things?

Citizens are obliged in conscience not to follow the directives of civil authorities when they are contrary to the demands of the moral order. We must obey God rather than men. (CCC 2256)

And the Catholic Church is only too happy to be the one to interpret and pass along God's demands.

The point is that any Catholic in public office, his first commitment has to be his faith. . . . (Bishop Thomas Tobin, November 23, 2009)

The Roman Catholic Church actually demands that Catholic judges obey Catholic law when deciding civil legal issues. (FFRF, Andrew Seidel)

Priorities: *The most tragic hunger and the most terrible anguish is not lack of food. It's much more about the absence of God and lack of true love, the love that was revealed to us on the cross.* (Cardinal Robert Sarah, May 22, 2011)

By the way, God comes before civil government not just for Catholics but also for the Baptists:

Whenever the civil government forbids the practice of things that God has commanded us to do, or tells us to do things He has commanded us not to do, then we are on solid ground in disobeying the government and rebelling against it. (Pat Robertson, 1984)

Civil government being ordained by God, it is the duty of Christians to render loyal obedience thereto in all things not contrary to the revealed word of God. . . . The state has no right to impose penalties for religious opinions of any kind. (Southern Baptist Convention 2005)

Arizona, Kansas, Louisiana, South Dakota, and Tennessee have passed laws that prohibit judges from considering *foreign laws* in their decisions. While this is typically understood to refer to Sharia law, it apparently could refer to any law from outside the US, including the Vatican City State. In Florida the proposed legislation included the words: *but not limited to, a foreign or international organization claiming the status of a country, state, or nation . . .*

Under this law could a judge reject a Catholic hospital's refusing to do an abortion based on *Catholic* law? Does religious freedom prevail over civil law?

Let every soul be subject unto the higher powers. For there is no power but of God: the powers that be are ordained of God. (Romans 13:1) In 2012, the Minnesota Catholic Conference, the church's political arm in the state, spent over half a million dollars to try to pass an amendment to the state constitution that would have supported church dogma and prevented gay marriages. It failed. We allow a foreign faith-based state all manner of privilege, and then see it attempt to amend our constitution. Clearly all Catholic entities in the US are subject to foreign control.

Many states allow children to be exempt from vaccinations because of religious concerns.

As of 2013, thirty-eight states and the District of Columbia still had religious exemptions for child abuse in their civil codes, and seventeen had had religious exemptions for felony crimes. (Paul A. Offit, M.D., Author of, *Bad Faith: When Religious Belief Undermines Modern Medicine*)

Because of Christian influence, only the United States and Somalia have not ratified the U.N. Convention on the Rights of the Child.

Teach Your Children . . . Well?

Public Education vs. Catholic Education

In a 1787 letter to James Madison, Thomas Jefferson urged his protégé to *educate and inform the whole mass of the people. . . . They are the only sure reliance for the preservation of our liberty.* (FFRF)

The people of Minnesota felt similarly and signaled that clearly in 1857 in the state's constitution, which not only supported public education but also prohibited aid to sectarian schools.

Section 1. Uniform system of public schools. The stability of a republican form of government depending mainly upon the intelligence of the people, it is the duty of the legislature to establish a general and uniform system of public schools. The legislature shall make such provisions by taxation or otherwise as will secure a thorough and efficient system of public schools throughout the state.

Section 2. Prohibition as to aiding sectarian school. In no case shall any public money or property be appropriated or used for the support of schools wherein the distinctive doctrines, creeds or tenets of any particular Christian or other religious sect are promulgated or taught. (Article XIII, Minnesota State Constitution, adopted on October 13, 1857, and generally revised on November 5, 1974)

Pope Pius XI felt otherwise about education. In 1939 he condemned public schools. He was not the first nor, unfortunately, the last to do so.

I imagine every Christian would agree that we need to remove humanism from the public schools. There is only one way to accomplish this: Abolish the public schools. (Robert L. Thoburn, 1986)

The Church's very mission is to defend and spread the truth. For this reason the Pope [Benedict XVI] has asked government leaders to continue to permit the Church freely to attend to the education of young people for the common good. (L'Osservatore Romano, December 14, 2012)

And, in the end, do not forget that the greatest legacy you can give your children is faith. (Pope Francis, VIS, 10 January 2016)

In fact, there is a determined effort by secularists to force Catholic Schools to tailor their strictures to the wisdom of civil law. (Bill Donohue, November 19, 2014)

Yes!

Sunday School should really not be called *school.* In a school one should learn real things. In school one should learn about reality, about rational thought and science.

Contrast this with Catholic parochial schools and religiously based *voucher* schools that are not supervised by the state but are funded by state taxpayers' money. They can teach anything —including religion, such as a *Qur'an-guided journey toward active global citizenship* (IRS) or instruction *presented from God's point of view*—and they don't require licensed teachers.

Think about what that teaching might include. As you do so, recall that at various points in its history the Catholic Church officially has not wanted people to:

believe in the heliocentric solar system
read the Bible in their own languages
learn scientifically valid sex education

know of Voltaire, Hume, Descartes, Pascal, Kant, Locke, Copernicus, and Kepler

read Edward Gibbons' *History of the Rise and Fall of the Roman Empire*

accept the Violence Against Women Act without *Conscience Protection* provisions

speak in church if they are female

accept reason independent of faith

report the crimes of rape and torture of children to civil legal authorities

obey civil authorities if they do not agree with the church

stop oppressing women

live without fear

continue with democracy

tolerate nonbelievers

prefer secular education

use birth control

have an abortion even if their lives depended on it

accept Jews (until 1964, at least)

know it is saving them from sins it invented

divorce

Our nation suffers the consequences of religiously set education agendas. Consider this statement in the A.C.E curriculum, a product of Accelerated Christian Education, Inc.:

The curriculum conveys Biblical values and concepts considered foundational to meaningful interpersonal relationships and productive learning in such a way as to become life-shaping influences. . . . Some may question, 'Isn't evolution just an innocent, harmless little theory?' Absolutely not! It is a lie. . . . It is plainly understood from the Word of God and from the current state of the world that the theory of evolution is extremely damaging to children individually and to society as a whole. This is true for two reasons. First of all, the theory of evolution denies the principle of the individual's accountability to our Almighty and Holy God. It thus erodes the individual's moral foundation. . . . Second, the theory of evolution totally removes the basis for self-worth in our children. . . . Many people today, it appears, think secular professors and doctors know more about educating children than does God! Many parents have become convinced that modern science and intellectualism require them to believe philosophies centered around secular humanism, such as the theory of evolution. . . . We must stand upon what God Himself has proclaimed to be truth! God is truth, and in Him is no deceit. (A.C.E)

Now the above might not seem threatening if it was the theology of a small local congregation. But what if it's the deeply held faith of half the country?

All schools, at all levels, should be teaching critical thinking skills to every student in the state, just as we teach mathematics and English at every level. It would be helpful in understanding mathematics, science, politics, religions and sales pitches of all kinds.

This, unfortunately, is not the policy of the Catholic Church.

Through the grace of the sacrament of marriage, parents receive the responsibility and privilege of evangelizing their children. Parents should initiate their children at an early age into the mysteries of the faith of which they are the 'first heralds' for their children. (CCC 2225)

Note: the Catholic Church made marriage a sacrament in 1562 at the Council of Trent.

Catholic parents also have the duty and right of choosing those means and institutions through which they can provide more suitably for the Catholic education of their children, according to local circumstances. (CCL 793.1)

Parents are to entrust their children to those schools which provide a Catholic education. If they are unable to do this, they are obliged to take care that suitable Catholic education is provided for their children outside the schools. (CCL 798)

Pope Benedict XVI expanded on this during his meeting with Catholic educators in Washington, D.C., in April 2008:

Teachers and administrators, whether in universities or schools, have the duty and privilege to ensure that students receive instruction in Catholic doctrine and practice. (Pope Benedict XVI)

Unfortunately, a majority of United States Representatives agree with giving foreign-controlled private schools a free rein to indoctrinate students as they wish. In writing about the new Every Student Succeeds Act (signed December 10, 2015), Congressman Kline stated, *. . . and the agreement prevents any federal interference in the nation's private schools and home schools.* (*Northfield News*, December 26, 2015)

No child should be denied his or her right to an education in faith, which in turn nurtures the soul of a nation. . . . From this perspective one can recognize that the contemporary 'crisis of truth' is rooted in a 'crisis of faith.' Only through faith can we freely give our assent to God's testimony and acknowledge him as the transcendent guarantor of the truth he reveals. . . . A particular responsibility therefore for each of you, and your colleagues, is to evoke among the young the desire for the act of faith, encouraging them to commit themselves to the ecclesial life that follows from this belief. It is here that freedom reaches the certainty of truth. (Pope Benedict XVI)

Who should pay for that Catholic indoctrination? According to Catholic Canon Law:

Parents also have the right to that assistance, to be furnished by civil society, which they need to secure the Catholic education of their children. (CCL 793.2)

Civil society pays for Catholic education? Remember that article in the Minnesota State Constitution? *In no case shall any public money or property be appropriated or used for the support of schools wherein the distinctive doctrines, creeds or tenets of any particular Christian or other religious sect are promulgated or taught.* The First Amendment to the US Constitution offers some obstacles too, specifically the part that *prohibits the making of any law respecting an establishment of religion.*

C is for Catechism . . .

It seems like the Catholic Church will have to take care of inculcating its values in its members on its own. To do that, it relies heavily on the Catechism of the Catholic Church, promulgated by Pope John Paul II in 1992. It *contains the fundamental Christian truths formulated in a way that facilitates their understanding.* (United States Conference of Catholic Bishops) Paradoxically, even as it undergirds orthodoxy, it appears to espouse following one's conscience.

A human being must always obey the certain judgment of his conscience. (CCC 1800)

Conscience is a judgment of reason whereby the human person recognizes the moral quality of a concrete act that he is going to perform, is in the process of performing, or has already completed. In all he says and does, man is obliged to follow faithfully what he knows to be just and right. . . . (CCC1778)

Man has the right to act in conscience and in freedom so as personally to make moral decisions. . . . (CCC 1782)

Is someone who in good conscience acts wrongly guilty in God's sight? No. (YOUCAT 298)

A person is capable of distinguishing good actions from bad ones because he possesses reason and a conscience, which enable him to make clear judgments. (YOUCAT 291)

But wait. Maybe that doesn't apply when your conscience is at odds with church doctrine?

Personal conscience and reason should not be set in opposition to the moral law or the Magisterium of the Church. (CCC 2039)

What is Faith? Faith is the power by which we assent to God, acknowledge his truth, and commit ourselves personally to him. (YOUCAT 307)

The disciple of Christ must not only keep the faith and live on it, but also profess it, confidently bear witness to it, and spread it. . . . (CCC 1816)

In September 2013, Pope Francis touched on some of these issues in a letter written to *a non-believer,* Eugenio Scalfari, the editor of

La Repubblica and a stalwart atheist. Below are two translations of a section of the pope's letter—two because the English versions convey slightly different meanings . . . and remind us how even today meanings can shift during the conversion of a papal quote from one language to another:

1) *Above all, you ask if the God of Christians forgives those who do not believe and who do not seek faith. Given the premise, and this is fundamental, that the mercy of God is limitless for those who turn to him with a sincere and contrite heart, the issue for the unbeliever lies in obeying his or her conscience. There is sin, even for those who have no faith, when conscience is not followed. Listening to and obeying conscience means deciding in the face of what is understood to be good or evil. It is on the basis of this choice that the goodness or evil of our actions is determined.* (Pope Francis, September 11, 2013)

2) *First of all, you ask if the God of the Christians forgives those who do not believe and do not seek faith. Given that—and this is fundamental—God's mercy has no limits if he who asks for mercy does so in contrition and with a sincere heart, the issue for those who do not believe in God is in obeying their own conscience. In fact, listening and obeying it, means deciding about what is perceived to be good or to be evil. The goodness or the wickedness of our behavior depends on this decision.* (Pope Francis, September 11, 2013)

The pope's letter was especially notable, said the Religion News Service, for *its open and honest assessment of the spiritual state of nonbelievers, and for seeming to open the door to the idea that notions of sin, conscience and forgiveness can be decided outside of the Catholic Church.* It was not the first (or probably the last) time he has weighed in on the subject of atheism.

The Lord has redeemed all of us, all of us, with the Blood of Christ: all of us, not just Catholics. Everyone! 'Father, the atheists?' Even the atheists. Everyone! (Pope Francis May 22, 2013)

Pope Paul VI suggested that believers sometimes contribute to atheism:

Hence believers can have more than a little to do with the birth of atheism. To the extent that they neglect their own training in the faith, or teach erroneous doctrine, or are deficient in their religious, moral or social life, they must be said to conceal rather than reveal the authentic face of God and religion. (December 7, 1965)

On the other hand, St. Augustine viewed persecuting heretics as doing them a favor, saving them from eternal damnation. Something has changed here.

The Catholic Church as Arbiter of Morals, Civil Law

One of the Catholic Church's most visible roles in recent decades has been in the social and political arenas. Let's look at some of the controversial issues in which it has involved itself.

Birth Control/Overpopulation

Sacred Scripture and the Church's traditional practice see in large families a sign of God's blessing and the parents' generosity. (CCC 2373)

Every family is a cell of society, but the large family is a richer, more vital cell, and the state has much to gain by investing in it. (Pope Francis, VIS, December 28, 2014)

Pope Francis continued with an offer to pray for those families in economic crisis or those where the parents have lost their jobs. The large family just has to have faith that everything will turn out okay. And the state should also support large families.

A month later—after much criticism—the pope expanded on his remarks: *Some believe that—excuse the expression—to be good Catholics we must be like rabbits. No. Responsible parenthood. This is clear, and for this reason in the Church here are matrimonial groups, experts in this. I know of many, many legitimate ways to achieve this. . . . On the other hand, for the poorest people, a child is a treasure. It is true, we must also be careful here. But for them, a child is a treasure. God knows how to help them.* (Pope Francis, VIS January 20, 2015)

But, do the parents? And does the state?

Nothing here is said about overpopulation. From the scientific perspective, *Overpopulation is the world's top environmental issue.* (SUNY College of Environmental Science and Forestry, April 20, 2009)

Educated people are leading the way, even against the Catholic Church's anti birth-control morality stand.

Advocates of population moderation cite issues like quality of life, carrying capacity and risk of starvation as being a basis to argue against high human population growth. . . . Most countries have no direct policy of limiting their birth rates, but the rates have still fallen due to educating people about family planning and increasing access to birth control and contraception. (Wikipedia)

Abortion

The following quotes of the Catholic Church define its position on human life.

God-given human life is God's own property. . . . ***Every child, from the moment of conception on, has a right to life*** *[bold emphasis added].* (YOUCAT 383)

The zygote is God's property, not the parent's property?

God alone is Lord over life and death. Except in the case of legitimate self-defense of oneself or another, no one may kill another human being. (YOUCAT 378)

God is in control of life and death, especially for his property—except in self-defense. Which contradicts the Most Reverend Leonard P. Blair, Bishop of Toledo, who said, according to the teaching of the church, *that it is not morally acceptable to destroy one human life, even in its embryonic stage, in order to save another human life.* (July 5, 2011)

He created human beings and let them develop according to the internal laws that he gave to each one so they would reach their fulfillment. (Pope Francis, VIS October 27, 2014)

In [God's] hand is the life of every living thing and the breath of all mankind. (Job 12:10, CCC 2318) When it comes to life, according to the Church and the Bible, God is the decider.

Human life is sacred because from its beginning it involves the creative action of God and it remains for ever in a special relationship with the Creator, who is its sole end. (CCC 2258)

This bears emphasizing: ***Human life is sacred because from its beginning it involves the creative action of God.***

The human being is made in the image and likeness of God. And we all carry within us the breath of life from God, and every human life, the Bible tells us, is under the special protection of God. (Pope Benedict XVI, Feb. 6, 2013)

Even a birth physical defect or severe mental handicap is made in the image and likeness of God?

Before I formed thee in the belly I knew thee; and before thou camest forth out of the womb I sanctified thee, and I ordained thee a prophet unto the nations. (Jeremiah 1:5)

He created us, He governs us, He nurtured us even in the womb, brought us into the world, and still supplies us with all the necessaries of life and maintenance. (CCT)

Well . . . almost.

I ask everyone to continue to pay special attention to this most important issue of respect for human life from the moment of conception . . . safeguarding every human being from the very first moment of his or her existence. (Pope Francis, May 12, 2013)

Everything that exists depends on God and continues in being only because God wills it to be. (YOUCAT 44) [Emphasis added]

Abortion, said Pope Francis, is a 'scientific problem because there is a human life there, and it is not lawful to take out a human life to solve a problem.' (Catholic League, November 18, 2014)

But, we celebrate the crucifixion of Jesus in every Catholic Mass. Was he not the Son of man (human) and was not the crucifixion intended to solve a problem?

He not only gives them being and existence, but also, and at every moment, upholds and sustains them in being, enables them to act and brings them to their final end. (CCC 301)

The Bible, the Catholic League, two Popes, and three different Catholic catechisms all agree: God creates life at conception and then owns it, nurtures it, is responsible for it and is responsible for ending it.

Every child, from the moment of conception on, has a right to life. *[Emphasis added]* (YOUCAT 383)

Let's see how God does. Let's examine some scientific data regarding pregnancy in God's natural world. In nature, 50 percent of all fertilized eggs are lost before a woman's missed menses. (UCSF Medical Center) Studies reveal that anywhere from 10 to 25 percent of all clinically recognized pregnancies end in miscarriage. Miscarriage is a term used for a pregnancy that ends on its own within the first 20 weeks of gestation. (American College of Obstetricians and Gynecologists)

This means nearly 75 percent of conceptions are lost or miscarry before the twentieth week. God is batting about .250 here. *At no point in time does something that he [God] has created fall out of his hands.* (YOUCAT 49) But, if conditions regarding the pregnancy are not suitable, and since the pregnancy is still in the hands of God, God ends it. God aborts it.

Science has even identified some of the ways that God misses: fertilized eggs that harbor chromosome abnormalities (most common cause), fertilized eggs that don't implant, blighted ovum, hormone problems, mom's health, infections, trauma to uterus, diabetes, age of mother and father. In all these cases and even in the cases where the mother might be somewhat responsible (e.g., the mother avoids available pre-natal medical care), we don't label the miscarriage as a murder. It happens. Again, for God, the end justifies the means.

Couldn't God simply have skipped the conceptions in the first place as a way of avoiding his acts of abortion? God certainly foresaw what was going to happen. He could have made it look natural, since not every sperm/egg encounter results in a fertilization. Some Catholic theologian needs to explain why God would allow a conception followed by a natural, thus God-driven, abortion.

What if external conditions are not suitable, such as rape, incest, accident, or a pregnancy not wanted for a myriad of social, economic, and physical reasons? Why should a baby be born into an environment where it is not wanted or its chance of a happy survival is nil? Shouldn't the mother have something to say here? Shouldn't she at least be allowed the option of following God's example? After all, God also had no trouble killing innocent embryos in Noah's flood (Genesis 7,8) and in Sodom and Gomorrah (Genesis 19:25) and at the Passover slaughter (Exodus 12:29), the Midianite attack (Numbers 31:17), or when Joshua leveled Jericho (Joshua 6:21). And, it might be easy to assume there may be many pregnant unsaved mothers at the end of the world.

The mother Church, like Jesus, teaches by example. . . . (Pope Francis, September 10, 2014)

God's action is the model for human action. (CCC 2172)

In *Roe v. Wade* (January 22, 1973), the Supreme Court of the United States deemed abortion a fundamental right under the United States Constitution. Again, the people via the court have decided and this time agree with the de facto Catholic position: Humans need to follow God's example. Thus, if things are not going correctly, abort.

Toward Reconciliation?

In a recent encyclical letter, *Laudato Si: On Care for Our Common Home*, Pope Francis indicated his agreement with scientific findings regarding pollution, climate change, water resources, and biodiversity, noting that:

Following a period of irrational confidence in progress and human abilities, some sectors of society are now adopting a more critical approach. We see increasing sensitivity to the environment and the need to protect nature, along with a growing concern, both genuine and distressing, for what is happening to our planet. Let us review, however cursorily, those questions which are troubling us today and which we can no longer sweep under the carpet. Our goal is not to amass information or to satisfy curiosity, but rather to become painfully aware, to dare to turn what is happening to the world into our own personal suffering and thus to discover what each of us can do about it.

Is this now the official Catholic definition of science?

The essence of the Enlightenment, from a Catholic perspective, is this:

Though faith is above reason, there can never be any real discrepancy between faith and reason. Since the same God who reveals mysteries and infuses faith has bestowed the light of reason on the human mind, God cannot deny himself, nor can truth ever contradict truth. Consequently, methodical research in all branches of knowledge, provided it is carried out in a truly scientific manner and does not override moral laws, can never conflict with the faith, because the things of the world and the things of faith derive from the same God. (CCC 159)

The pope is looking for a dialog with science that is open and free of preconceptions.

Therefore, it seems very positive, both for us and for the society in which we live, to pause and discuss a reality as significant as faith, which points us to the teachings and person of Jesus . . . throughout the centuries of modernity, a paradox was witnessed: the Christian faith, whose newness and influence on humanity were expressed by the symbol of light, has been often characterized as the darkness of superstition in opposition to the light of reason. Thus, between the Church and Christian inspired culture on the one hand, and modern culture shaped by the Enlightenment on the other, a point was reached where there was no longer any dialogue. The time has now finally come, ushered in by the Second Vatican Council, for a dialogue that is open and free of preconceptions, and which reopens the doors to a responsible and fruitful encounter. (Pope Francis, Letter to a Non-Believer, September 4, 2013)

Uneducated is Better?

The above provides more evidence, it seems, that enterprises, like other organisms, do evolve, even enterprises as large as the Catholic Church. Change comes slowly, however, and faces resistance. Even as the pope points to the critical importance of learning, especially higher learning, to the future of our planet, others are suggesting that it is education that threatens Catholicism if not all Christianity:

It is precisely the 'better educated' class who are leading the militant secularist crusade. They are the ones who seek to ban the public expression of religion, make movies that insult Catholicism, paint priests as molesters, ban Bibles from hotels, display vile anti-Catholic billboards, attempt to crash the St. Patrick's Day parade, sponsor Catholic-bashing artistic exhibitions, force Catholic non-profits to pay for abortion-inducing drugs, and the like. (Bill Donohue, Catholic League, March 4, 2014)

So, Bill, does that mean those who are *worse educated* are your sectarian Christian/Catholic partners? You seem to infer, at least, that *better educated* is bad when everyone else thinks *better educated* should be good. Obviously you disassociate yourself from the ranks of the *better educated*—even though you claim to have a PhD.

George F. Will in a *Washington Post* article (about baseball) of January 28, 2015, wrote . . . *the intelligentsia, which usually is the last to understand things . . .*

A review of George's education and experience—he claims a BA from Trinity College, a BA and MA from Magdalen College, Oxford, and MA and PhD degrees from Princeton University—might lead one to place him in the intelligentsia class, i.e., a social class of people engaged in complex mental labor aimed at guiding or critiquing, or otherwise playing a leadership role in shaping a society's culture and politics. (Sss Wikipedia)

If that is the case, it means it is likely that he, in his own words, *is the last to understand things.* So why is he still writing for the *Post*?

On the other hand, if George feels he is not in the intelligentsia class, then why is George still writing for the *Post*?

In fact, the notion that more education tends to make one less religious and less superstitious is reinforced in studies reported by Naci Mosan and Luiza Pogorelova via the National Bureau of Economic Research in *The Economist*, October 11, 2014. While this may explain why Bill and George view education and intelligence as a negative—and view those who employ them as suitable for ad hominem attacks—it's hard to find arguments that support the value of ignorance. Ignorance is *not* bliss, and what you don't know *can* hurt you. In spite of what Bill and George and others say, education, rationality, and intelligence are still positives.

A question one might ask is: How could Christianity or Islam grow to acquire over a billion followers each? These numbers alone must argue for legitimate, sacred foundations to these religions. In the next chapter we will explore and compare a religion to an analogous secular socio-economic phenomenon.

Part VI

Baseball and Catholicism

If God let you hit a home run last time up,
then who struck you out on the time before that?
(Sparky Anderson)

What do professional baseball and the Catholic Church have in common? A lot, it seems.

For one thing, they both attract large numbers of people. Major League Baseball (MLB) has the highest overall attendance of professional sports in the world—more than 74 million in 2013. The Catholic Church in the United States, with more than 69 million members, is the largest religious denomination in the United States, comprising 22 percent of the population.

For another thing, both of these enterprises give rise to the same question: How does something become so important in so many people's lives? One is supported by a god, which may give it an advantage, and has had 2,000 years to develop, but the other has managed to do nicely without a god and in only 200 years.

The idea of the similarities between baseball and the church is not original to me. In 1988 the movie *Bull Durham* began with these words:

I believe in the Church of Baseball. I've tried all the major religions and most of the minor ones. I've worshipped Buddha, Allah, Brahma, Vishnu, Siva, trees, mushrooms, and Isadora Duncan. I know things. For instance, there are 108 beads in a Catholic rosary and there are 108 stitches in a baseball. When I learned that, I gave Jesus a chance. But it just didn't work out between us. The Lord laid too much guilt on me. I prefer metaphysics to theology. . . . [T]he only church that truly feeds the soul day in, day out, is the Church of Baseball.

And in a 2007 case centered on fantasy baseball, US Judge Morris Arnold, 8th US Circuit Court of Appeals, observed that *MLB is like a public religion.*

Need more evidence? Look at some comparisons. As you do, consider whether the evolution of baseball might help us understand the development of the Catholic Church, especially its earlier stages.

Similarities

ORIGINS

Baseball—Murky origins
Catholicism—Murky origins

Baseball—Evolved from multiple earlier games
Catholicism—Evolved from multiple earlier religions

Baseball—Of the top 25 famous baseball narratives, it's hard to tell which, if any, are true.

Catholicism—Of the top 25 famous Catholic narratives, it's hard to tell which, if any, are true.

Baseball—There was no very first baseball game.
Catholicism—There was no very first Catholic mass.

Baseball—Existed long before written rules appeared
Catholicism—Existed long before written rules appeared

Baseball—Though it now has its official rule book, rules are still in flux.

Catholicism—Though it now has its official catechism, rules are still in flux.

Baseball—The rules have evolved over the years: In the game long ago called *Base Ball*, a player could not steal a base, could not use a glove, and pitching was underhanded.

Catholicism—The rules have evolved over the years: Long ago priests could get married; there was no pope, only bishops; masturbation was a mortal sin; and Mass was in Latin.

STRUCTURE

Baseball—A family-oriented activity in which allegiance frequently is determined by location and/or family tradition

Catholicism—A family-oriented activity in which allegiance frequently is determined by location and/or family tradition

Baseball—Hierarchical form of administration: managers, owners, and commissioner, who is elected by the owners (a monarchy)

Catholicism—Hierarchical form of administration: bishops, cardinals, and the pope, who is elected by the cardinals (a monarchy)

Baseball—Has had *formations*: formation of the Union Association, Players' League, American Association

Catholicism—Has had *formations*: formation of the Lutheran Church, Church of England, Eastern Orthodox Church

Baseball—Men have all the starring roles; the environment is generally misogynistic, emphasizes manliness.

Catholicism—Men have all the starring roles; the environment is generally misogynistic, emphasizes manliness.

Baseball—Few significant roles for women, except as fans

Catholicism—Few significant roles for women, except as fans

Baseball—Makes use of neophyte helpers (bat boys and girls)

Catholicism—Makes use of neophyte helpers (altar server boys and girls)

Baseball—Participants (players, managers, owners) can be banned for breaking the rules.

Catholicism—Participants (members, priests, bishops) can be excommunicated for breaking the rules.

Baseball—Exempt from federal antitrust regulations

Catholicism—Exempt from federal antitrust regulations

Baseball—Activities subsidized by local, state, and federal governments

Catholicism—Activities subsidized by local, state, and federal governments

Baseball—Nearly every city has a baseball field.

Catholicism—Nearly every city has a Catholic church.

Baseball—Major venues (stadiums) upon first encounter take your breath away, induce euphoria, awe.

Catholicism—Major venues (cathedrals) upon first encounter take your breath away, induce euphoria, awe.

Baseball—Makes you feel important, part of something big

Catholicism—Makes you feel important, part of something big

ACTIVITIES / RITUALS

Baseball—Has *sacred* space (the playing field)

Catholicism—Has *sacred* space (the sanctuary)

Baseball—Practitioners wear uniforms: jerseys.

Catholicism—Practitioners wear uniforms: vestments.

Baseball—Has fanatical followers

Catholicism—Has fanatical followers

Baseball—Followers (fans) proclaim their allegiance by wearing team hats or shirts.

Catholicism—Followers (members) proclaim their allegiance by wearing crucifixes or religious medals.

Baseball—Public ceremonies (games) follow a format that is repeated on all occasions.

Catholicism—Public ceremonies (mass) follow a format that is repeated on all occasions.

Baseball—Events (games) have a principal actor, the pitcher.

Catholicism—Events (mass) have a principal actor, the priest.

Baseball—Except for the singing of an occasional song, baseball is not a fan-participation activity.

Catholicism—Except for the singing of an occasional song, Mass is not a member-participation activity.

Baseball—An activity (a bench-clearing brawl that results in injuries) is allowed during a game that would be illegal or considered unsavory on the street.

Catholicism—An activity (drinking *human blood*) is allowed during a Mass that would be illegal or considered unsavory on the street.

Baseball—A *sacrifice* can play a major role in determining who wins.

Catholicism—A *sacrifice* can play a major role in determining who goes to heaven.

Baseball—Events are made livelier by organ music.

Catholicism—Events are made livelier by organ music.

Baseball—You can't bring your own wine to events.

Catholicism—You can't bring your own wine to events.

Baseball—Rituals: Stand and sing the National Anthem, Seventh Inning Stretch

Catholicism—Rituals: Stand when the Bible is read, genuflect when entering pew

Baseball—Special events promote attendance: Bobblehead Day, Bat Day.

Catholicism—Special events promote attendance: Ash Wednesday, Blessing of Pets.

Baseball—Each playing season ends with a grand finale, the World Series.

Catholicism—Each liturgical season ends with a grand finale, Easter.

Baseball—Fans wish to be part of a virtual group of the winners, the successful ones.

Catholicism—Members wish to be part of a virtual group of the saved, the successful ones.

Baseball—Winning (the World Series) is the ultimate goal; only a few achieve it.

Catholicism—Winning (salvation) is the ultimate goal; only a few achieve it.

Baseball—*Out of the park* extends forever.

Catholicism—*Out of this world* extends forever.

SCOPE / INFLUENCE

Baseball—Followed by millions

Catholicism—Followed by millions

Baseball—Its popularity is promoted as proof of its status as *the National Pastime.*

Catholicism—Its popularity is promoted as proof of the status of the US as a *Christian Nation*.

Baseball—Starts training players and fans young before they can lose interest: few players make the majors

Catholicism—Starts training priests and members young before they can lose interest: few members make it to heaven

Baseball—Sends scouts across the oceans to promote itself, recruit new players and fans

Catholicism—Sends missionaries across the oceans to promote itself, recruit new priests and members

Baseball—Makes use of radio, TV to evangelize

Catholicism—Makes use of radio, TV to evangelize

Baseball—As a youngster you must often at least fake interest in baseball to appear normal.

Catholicism—As a youngster you must often at least fake interest in religion to appear normal.

Baseball—Common question: *You a Cubs fan?*

Catholicism—Common question: *Where do you go to church?*

Baseball—Circulates collectible baseball cards

Catholicism—Circulates collectible holy cards

Baseball—Has persons and objects of veneration: Hall of Fame in Cooperstown honors god-like heroes like Lou Gehrig, Jackie Robinson, and Babe Ruth and displays relics—balls, bats, uniforms, etc.

Catholicism—Has persons and objects of veneration, god-like saints like Francis of Assisi, Joseph, and John the Baptist, and displays relics—bones, hair, vestments, etc.

Baseball—Fans make pilgrimages to famous stadiums: Wrigley Field, Miller Park, Fenway Park.

Catholicism—Members make pilgrimages to holy shrines: Sistine Chapel, Virgin of Guadalupe, Lourdes.

VALUES

Baseball—Stresses self-control, discipline, and obedience
Catholicism—Stresses self-control, discipline, and obedience

Baseball—Stresses things that are fair and foul
Catholicism—Stresses things that are fair and foul

Baseball—Decrees winners and losers: the goal is to be a winner and ultimately to be a champion

Catholicism—Decrees what is good and bad: the goal is to be good and ultimately to be saved, to be with God

Baseball—Rules determine who is *safe* and who is *out.*
Catholicism—Rules determine who is *saved* and who is *out.*

Baseball—Has a referee
Catholicism—Has a confessor

Baseball—The presence of umpires encourages players to act ethically.

Catholicism—The presence of confessors encourages members to act ethically.

Baseball—Promotes fealty (to a team or city)
Catholicism—Promotes fealty (to God)

Baseball—Purports to be moral; ostensibly promotes values like cooperation, fairness, honesty, restraint, following the rules

Catholicism—Purports to be moral; ostensibly promotes values like cooperation, fairness, honesty, restraint, following the rules

Baseball—Has gambling, drug, and domestic abuse scandals
Catholicism—Has corruption, sex, and sex abuse scandals

Baseball—Much talk of *miracles*
Catholicism—Much talk of *miracles*

Baseball announcers and Catholic priests often employ very similar kinds of theology/logic.

Announcer:
Team A will need some big plays today to beat team B.—Easy to say.
If *Team A* wins, one can easily give credit to some big plays.
If *Team A* loses, one can easily say there were not enough big plays.

Priest:
Person A should pray to God to save Child B from death.—Easy to say.
If *Child B* survives, the prayer gets credit.
If *Child B* dies, the prayer was not prayed intensely enough.

This is actually official Catholic thinking: *Often we say: I have prayed, but it did not help at all. Maybe we are not praying Intensely enough.* (YOUCAT 507)

Baseball—True believers can only attempt to understand non-believers.

Catholicism—True believers can only attempt to understand non-believers.

Baseball—Fosters crowd behavior
Catholicism—Fosters crowd behavior

Baseball—An issue must be extremely serious for criticism of it to be tolerated.
Catholicism—An issue must be extremely serious for criticism of it to be tolerated.

Christianity (especially Catholicism) likes to trace its roots all the way back to God, even to the sixth day of the Creation. If baseball has all these things in common with it, then might baseball likewise be justified in tracing its origins back to God's creation?

Differences

There are significant differences as well between baseball and Catholicism.

Baseball—Payment is made before entering the stadium and it is required.
Catholicism—Payment is made after entering church and it is voluntary.

Baseball—In baseball you can not only steal (a base), but by doing so win the game and become a hero.
Catholicism –Stealing is not allowed, ever. It violates a commandment and qualifies you for eternal damnation.

Baseball—You always know who the winners (and losers) are.
Catholicism—You never know who the winners (and losers) are.

Baseball—Had tax-exempt status until giving it up in 2007
Catholicism—Still has tax-exempt status

Baseball—Arguing with others that one's team is best can be socially entertaining.

Catholicism—Arguing with others that one's church/religion is best can be socially divisive.

The World in 4015

Is sports replacing religion? Churches once *owned* Sundays. Then, between 1880 and 1920, through fights in legislatures and the courts against *blue laws* designed to restrict or ban some or all commercial activities on Sunday, baseball *got* Sunday afternoons. It eventually broke free of all Sunday restrictions. In the meantime, Catholic churches adjusted, scheduling Mass at times other than Sunday morning, like Saturday and Sunday evenings, when there would be fewer conflicts with sport contests of various sorts.

As the influence of religions on daily life has declined, the influence of sports has increased. Baseball came to be viewed as a contributor to fitness and as a positive moral influence, something that built character and minimized drunkenness, delinquency, and evildoing.

Will baseball last another 2,000 years? If so, what might it look like? Do you think you would recognize it? What new rules would it have? What new doctrines? In another 150 years, will people wonder if Babe Ruth was both human and God? And what will the people of 4015 know (or think they know) about the early days of baseball?

There's no guarantee that baseball will endure, of course—or even life on our planet, for that matter. Although attendance at professional baseball games has doubled in the last half century, its popularity could be eclipsed by another sport, e.g., football or soccer, just as Mithraism was overtaken by Christianity. (Then again, football could find itself sidelined by legal and economic woes: the NFL paid $765 million recently to settle a lawsuit over covered-up data relating to concussions. Or it might be so transformed by the physical dangers and liability issues that people lose interest in football altogether.)

There's no guarantee the Catholic Church will endure either, of course. For a variety of reasons, attendance at Catholic elementary-secondary schools in the US has halved in the last half century. It also faces economic challenges: In recent years, the church has paid an estimated $3 billion to victims relating to sex abuse by clergy. (Reuters, November 10, 2015)

So . . . what will be the major ways of viewing the world in 50 or 100 years? Will the majority of people still see the world through the lens of their faith? Or will they rely instead on what they've learned in school and on the scientific method's techniques for investigating phenomena, acquiring new knowledge, and correcting and integrating previous knowledge? For the sake of those who follow—my grandchildren, their children, and the other peoples of the world—I hope it's science and its dependence on evidence that wins the day.

Glossary

Agnostic One who accepts that it is impossible to know if a deity exists

Altruism Individual sacrifices made for the greater good of the population

Apostate One who rejected and left a religion

Atheism Absence of a belief in the existence of a deity

Bastard Person born of unmarried parents

Blasphemy An act held to be disrespectful toward a deity, religion, or belief

Calvinball A game in which the rules change during the course of play

Consequentialism A belief that the good ends do justify the means

Creationism Belief that all living creatures were created within a short time by a deity

Cult A small religion repudiated by the mainstream society

Deism Belief in an uninvolved deity based on reason, not religious dogma

Deity A supreme being, a god

Dominionism A Christian theocracy

Enterprise An activity with a goal

Evangelize Influence others by teaching, to convert them to your belief

Faith A belief with no evidence to support it

Free will The ability to act on your own, not limited or controlled by anything

Fundamentalist One who adheres to strict theological doctrines

Hearsay Unverified information heard or received from another

Heresy Any opinion opposed to official view or doctrine

Homiletics The art of preaching

Hypothesis A statement with the capability of being proven false

Justification Reunited with God, saved

Law An interesting pattern in the scientific data, with no known exceptions

Literalist One who adheres to the exact word

Martyr One who suffers or dies for a deity on the basis of faith

Miracle An event not consistent with known scientific laws of nature

Mystery Something impossible to explain in the rational natural world of science

Nihilist One who believes there is no God and therefore life is meaningless

Parochial Narrowly restricted in scope or outlook

Proselytize Preaching with the intent of converting others to one's own belief

Rational Using reason with its logic to arrive at a conclusion

Revelation Information made known by a deity

Schism Division of a group into factions

Science Knowledge gained by evidence via observation, experiment, and logic

Sectarian Supporting a particular religious belief exclusively

Secular Independent of and not connected or relating to any religion

Supernatural Beyond science and the laws of nature

Synod A meeting of church leaders to facilitate an incremental evolution of doctrine

Theist One who accepts the existence of a deity or deities

Theocracy A government controlled by a religion

Theodicy Attempts to resolve evil and suffering with benevolence of a deity

Theology Study of religious questions and the nature of a deity

Theory A rational explanation of why something happens

Xenophobia Fear of that which is foreign or strange

Famous Atheists

John Adams, Jessica Ahlquist, Alan Alda, Ethan Allen, Seth Andrews, Natalie Angier, Susan B. Anthony, Isaac Asimov, Paula Aspell, David Attenborough, Margaret Atwood

Dan Barker, Clara Barton, Lauren Becker, Ophelia Benson, Ingmar Bergman, Jamila Bey, Lewis Black, Napoleon Bonaparte, Julian Bond, Marsha Botzer, Amanda Brown, Ann Brusseel, Warren Buffet

George Carlin, Andrew Carnegie, Liz Cavell, Charlie Chaplin, Piper Chapman, Soraya Chemaly, Greta Christina, Arthur C. Clarke, George Clooney, John Conway, Laura Cooper, R. Elizabeth Cornwell, Adam Corolla, Jerry Coyne, Francis Crick, Marie Curie

Heina Dadabhoy, Clare Daly, Clarence Darrow, Richard Dawkins, Simone deBeauvoir, Catherine Deveny, Cameron Diaz, Marlene Dietrich, Margaret Downey, W.E.B. DuBois

Thomas Edison, Albert Einstein, Ralph Waldo Emerson

Richard Feynman, W.C. Fields, Carrie Fisher, David Fitzgerald, Harrison Ford, Jody Foster, Morgan Freeman, Sigmund Freud, Betty Friedan, Robert Frost, Stephen Fry

Helen H. Gardener, Vyckie Garrison, Bill Gates, Annie Laurie Gaylor, Lynnea Glasser, Debbie Goddard, Emma Goldman, Rebecca Goldstein, A.C. Grayling, Laci Green, Ruth Hurmence Green, D.J. Grothe, Helena Guzik

Edmund Halley, Sarah Haider, Rebecca Hale, Tracy Harris, Jennifer Hecht, Ernest Hemingway, Melody Hensley, Katharine Hepburn, Peter Higgs, Ayaan Hirsi Ali, Alfred Hitchcock, Christopher Hitchens, Alex Honnold

Robert G. Ingersoll

Roger Scott Jackson, Susan Jacoby, Mick Jagger, Billy Joel, Ellen Johnson, Angelina Jolie, Zinnia Jones, Scott Joplin

Helen Keller, Larry King, Paula Kirby, Keira Knightly, Joan Konner, Lawrence M. Krauss, Stanley Kubrick, Paul Kurtz

Nica Lalli, Burt Lancaster, Cloris Leachman, John Lennon, Leighann Lord, Liz Lutgendorff

Teresa MacBain, James Madison, Shirley Manson, Christopher Marlowe, Karl Marx, Jennifer McCreight, Thelma Butterfly McQueen, Anthea McTiernan, Hemant Mehta, John Stuart Mill, Arthur Miller, Ashley Miller, Maggie Miller, Moliere, Wolfgang Amadeus Mozart, D.M. Murdock, PZ Myers

Maryam Namazie, Taslima Nasrin, Ingred Newkirk, Randy Newman, Jack Nicholson, Diana Nyad, Bill Nye

Joyce Carol Oates, Dara Ó Briain, Eugene O'Neill, George Orwell

Thomas Paine, Sara Paretsky, Pragna Patel, Jen Peeples, Julia Phillips, Steven Pinker, Brad Pitt, Katha Pollitt, Carolyn Porco, Cole Porter, Paula Poundstone

Aron Ra, Christana Rad, James Randi, Ron Reagan, Christopher Reeve, Keanu Reeves, Gene Roddenberry, Edwina Rogers, Richard Rogers, Ray Romano, Andy Rooney, Sir Bertrand Russell

Carl Sagan, Margaret Sanger, Nina Sankari, Charles Schultz, Eugenie Scott, Shelley Segal, Farhana Shakir, George Bernard Shaw, Mary Ellen Sikes, David Silverman, Upton Sinclair, Kate Smurthwaite, Ailbhe Smyth, Jill Sobule, George Soros, Benjamin Spock, Elizabeth Cady Stanton, Denise Stapley, Gloria Steinem, Victor Stenger, Leonard Susskind, Stephanie Svan, Julia Sweeney, Jonathan Swift

Valerie Tarico, Teller, Mandisa Thomas, Pat Tillman, Linus Torvalds, Miranda Tradewell, Alan Turing, Mark Twain, Neil DeGrasse Tyson

Rebecca Vitsmun, Voltaire, Indre Viskontas

Barbara G. Walker, Ann Marie Waters, Ayanna Watson, James Watson, Rebecca Watson, H.G. Wells, Oscar Wilde, Ted Williams, Marlene Winell, Virginia Woolf, Frances Wright

Famous Theists

Mahmoud Ahmadinejad, Henry Alford, Samuel Alito, Leith Anderson, Karen Armstrong, Francis Asbury, Thomas Aquinas, Augustine

Michele Bachmann, Jim Bakker, Tammy Fay Bakker, Ann Bancroft, Paul Barnes, Ethel Barrymore, David Barton, Glen Beck, Michael Behe, Benedict XVI, Osama bin Laden, Tony Blair, Napoleon Bonaparte, Bono, Sonny Bono, Pat Boone, Sir Robert Boyle, John Brooks, Gary Busey, George W. Bush

John Calvin, Harold Camping, Gretchen Carlson, Jimmy Carter, Peter Cartwright, George Washington Carver, Johnny Cash, Saint Catherine, G.K. Chesterton, Jesus Christ, Titus Coan, Douglas Coe, Francis Collins, Charles Colson, Ray Comfort, Sean Connery, Dane Cook, Kenneth Copeland, Nicolas Copernicus, Luis Cortes, Ann Coulter, Larry Craig, William Lane Craig, Bing Crosby, Dan Crouch, Paul Crouch

Charles Darwin, Leonardo da Vinci, James Dobson, Creflo Dollar, Dorothy Day, Rene' Descartes, Bill Donahue, Fyodor Dostoevsky, Faye Dunaway, Albrecht Durer

Stuart Epperson

Jerry Falwell, Mia Farrow, Henry Emmerson Fosdick, Lonnie Frisbee

Clark Gable, Galileo Galilei, Michael Gerson, Mel Gibson, Douglas Goodman, Billy Graham, Franklin Graham, Oliver Green

John Hagee, Matthew Hagee, Ted Haggard, Ken Ham, Sean Hannity, Helen Hayes, Susan Hayward, Saint Hildegarde, Paul Jennings Hill, Benny Hinn, Adolf Hitler, John Hollingshead, Bob Hope, Kent Hovind, L. Ron Hubbard, Mike Huckabee, Johan Huibers, Whitney Houston, Bill Hybels

Laura Ingraham

T.D. Jakes, King James, Joan of Arc, Bob Jones, Jim Jones

Grace Kelly, Anthony M. Kennedy, D. James Kennedy, Johannes Kepler, Martin Luther King, Jr., Diane Knippers

Tim LaHaye, Beverly LaHaye, Richard Land, Gottfried Wilhelm von Leibniz, C.S. Lewis, Alphonsus Liguori, James Linzey, Martin Luther, Erwin Lutzer, Barry Lynn

James Clerk Maxwell, Josh McDowell, Joyce Meyer, Michelangelo, William Miller, Dwight L. Moody, Beth Moore, Thomas More, Terry Mortenson, Muhammad

Richard John Newhaus, Sir Issac Newton, Mark Noll, Chuck Norris

Bill O'Reilly, Joel Osteen

J. I. Packer, Sarah Palin, Blaise Pascal, Norman Vincent Peale, Tony Perkins, Fred Phelps, Pablo Picasso, James Pike, Peter Popoff, Carrie Prejean, Elvis Presley

Ralph Reed, Vaughn Reeves, Mary Lou Retton, Oral Roberts, Pat Robertson, Ida Robinson, Fred Rogers, Mitt Romney

Rick Santorum, Jack Schaap, Rob Schenck, Robert H. Schuller, Jay Sekulow, Elizabeth Ann Seton, Fulton Sheen, John Smyth, Stephen Strang, Billy Sunday, Jimmy Swaggart

Tim Tebow, Mother Teresa, Randall Terry, Robert Tilton, Donald Trump

Thomas Ware, Mike Warnke, Rick Warren, John Wayne, John Wesley, Kanye West, James Springer White, Carl Wieland, Gerald Winrod, Ralph Winter, Jeremiah Wright, John Wycliffe

Brigham Young

Suggested Reading

Books

Catechism of the Catholic Church
Revised in accordance with the official Latin text, promulgated by Pope John Paul II

Deconverted: A Journey from Religion to Reason—Seth Andrews
Takes readers on a deeply personal journey of coming out

God Is Not Great: How Religion Poisons Everything—Christopher Hitchens
Hitchens tells the personal story of his own dangerous encounters with religion and describes his intellectual journey toward a secular view of life based on science and reason

New Commentary on the Code of Canon Law—Edited by Beal, Coriden, Green
Commissioned by The Canon Law Society of America, English Study Edition

The Age of Reason: Being an Investigation of True and Fabulous Theology—Thomas Paine
The most formidable weapon against errors of every kind is Reason.

The Companion to the Catechism of the Catholic Church
A compendium of texts referred to in the Catechism of the Catholic Church

The Faith of 50 Million: Baseball, Religion, and American Culture
A collection of essays edited by Christopher H. Evans and William R. Herzog II

The Greatest Show on Earth: The Evidence for Evolution—Richard Dawkins
A summary of the evidence that evolution is actually a fact

Why Evolution is True—Jerry A. Coyne
No subject has caused more fascination and fury than evolution

YOUCAT (Youth Catechism of the Catholic Church)
Shows what the Catholic Church believes today

On the Web

http://www.catholiccrossreference.com/catechism
Catechism reference

https://www.biblegateway.com/passage
Reference to many bibles

http://www.vis.va/vissolr/index.php?lang=en
Vatican Information Service

http://atheists.org/page.aspx?pid=329
Fighting for the civil liberties of atheists

http://ncse.com/about
The National Center for Science Education

http://ffrf.org
Freedom from Religion Foundation (Church & State separation)

Blogs

http://rosarubicondior.blogspot.com
Religion, Science and Politics for the Humanist

http://freethoughtblogs.com
Reason, Discussion, Opinion

http://www.patheos.com
Conversation on Faiths and Atheism

Postscript

Is God willing to prevent evil, but not able?
Then he is not omnipotent.

Is he able to prevent evil, but not willing?
Then he is not omnibenevolent.

Is he both able and willing?
Then why does evil exist?

Is he neither able nor willing?
Then why call him God?

(Epicurus 341–270 BCE)

Lucky
One
Making It Past Polio and Despair
Richard Maus